Soil Testing: Correlating and Interpreting the Analytical Results

ASA Special Publication Number 29

Proceedings of a symposium sponsored by Divisions S-4, S-8, and A-4 and the Soil Testing and Plant Analysis Committee of the American Society of Agronomy, Crop Science Society of America, and Soil Science Society of America. The papers were presented during the annual meetings in Knoxville, Tennessee, August, 1975.

Editorial Committee

T. R. Peck, Chairman
J. T. Cope, Jr.
D. A. Whitney

Organizing Committee

C. D. Welch, Chairman
T. R. Peck
E. C. Doll
V. A. Bandel
W. D. Pardee

Editor-in-Chief: **Matthias Stelly**
Managing Editor: **David M. Kral**
Assistant Editors: **Linda C. Eisele**
Judith H. Nauseef

1977

Reprinted: 1977
1978
1980
1984

Published by the

AMERICAN SOCIETY OF AGRONOMY
CROP SCIENCE SOCIETY OF AMERICA
SOIL SCIENCE SOCIETY OF AMERICA
677 South Segoe Road
Madison, Wisconsin 53711

American Society of Agronomy
Crop Science Society of America
Soil Science Society of America
677 South Segoe Road, Madison, Wisconsin 53711 USA

Library of Congress Catalog Card Number: 76-27889
Standard Book Number: 0-89118-047-8

Printed in the United States of America

Contents

Foreword

The World's population increases and our knowledge of the limits of our finite resources become ever clearer. Thus, it becomes increasingly important that those of us engaged in research in the production of the food and fiber needs of mankind provide not just information which can provide for maximum production but rather we must provide that information which can lead to production maximization in terms of the most efficient use of our resources.

Soil testing has been a valuable agronomic practice for many years. However, it is becoming increasingly important if we are to realize the desired, indeed essential, highly efficient production of food and fiber. This special publication presents information aimed at this increased efficiency through improved correlation and interpretation of the analytical results obtained during soil testing.

The papers included in this special publication were presented at a symposium held August 27, 1975 during the annual meetings of the American Society of Agronomy, Crop Science Society of America, and Soil Science Society of America at Knoxville, Tennessee. The symposium was cosponsored by three divisions of the societies and one committee—Soil Fertility and Plant Nutrition (S-4), Fertilizer Technology and Use (S-8), Extension Education (A-4), and the Soil Testing and Plant Analysis Committee (S-877).

Our appreciation goes to C. D. Welch, T. R. Peck, E. C. Doll, V. A. Bandel, and W. D. Pardee who organized the symposium; to the editors of this special publication T. R. Peck, J. T. Cope, and D. A. Whitney; and to the authors of these timely and useful papers.

F. L. Patterson, President
American Society of Agronomy
R. L. Mitchell, President
Crop Science Society of America
C. B. Davey, President
Soil Science Society of America

Preface

This publication emanates from a special symposium presented at the 1975 American Society of Agronomy annual meeting in Knoxville, Tennessee. The objectives of the symposium were to present procedures and philosophies used to develop guides for soil liming and fertilization practices based on soil test results that result in improved plant growth. This publication is designed to cover topics either omitted or covered inadequately in the book *Soil Testing and Plant Analysis*. The authors were invited because of expertise in their respective areas. The authors present a range of views on the central theme giving the reader exposure to different philosophies on soil test correlation and interpretation.

Correlation and interpretation are two-stages of developing a soil testing program. Correlation is the process of relating plant performance to an analytical result. Frequently limitations in personnel and budget restrict the experimental work to less than the full range of field conditions. Interpretation is the modifying process to broaden the correlation over a wider range of field conditions. Discussions of the two processes are interwoven in the papers. Presentations by Melsted and Nelson deal more completely with correlation while the major theme of the other papers is interpretation.

The American Society of Agronomy 1975 Soil Testing and Plant Analysis Committee composed of R. J. Bartlett, J. B. Jones, T. R. Peck, E. O. McLean, J. E. Sedberry, and C. D. Welch, chairman organized the program. Special appreciation is expressed to E. C. Doll and C. B. Davey for initiating the idea to schedule the program.

T. R. Peck
J. T. Cope, Jr.
D. A. Whitney
Editorial Committee

The Mitscherlich-Bray Growth Function

S. W. Melsted and T. R. Peck[1]

Agronomists are concerned in many ways with the growth of plants. Soil testing, as a part of the field of agronomy, is concerned with the relationships between the quantity and chemical form of the essential nutrients in the soil and the resulting plant growth. To conduct good soil test correlation studies it is essential that experiments be designed in such a way that the results may be interpreted in terms of known crop response patterns. Three general types of mathematical expressions are frequently used to relate soil nutrient levels to crop growth, namely the exponential, the quadratic, and the straight line.

Immobile nutrients, such as P, K, Ca, and Mg that are absorbed by the soil and, therefore, diffuse, migrate, and move at a much slower rate than the root tips penetrate the soil, can usually be related to growth through either the exponential or quadratic expression. Mobile nutrients such as nitrates and borates that are not absorbed by the soil and can diffuse, migrate, and move in and with the soil water at rates much faster than the root tips advance, are more often related to plant growth through a straight line function.

Many, if not most, biologic reactions are exponential in character. The development of a plant follows an exponential relationship with time, as its growth progresses through the growing season, giving rise to the conventional sigmoid-shaped growth curve. Many growth factors can be shown to have an exponential relationship to yields. Therefore, biologically an exponential function has merit over the quadratic function in attempting to relate a growth factor mathematically to yield responses, or predicting responses from experimental data. A review of how plants grow and examples of the applicability of the Mitscherlich-Bray growth function to soil test correlation studies will be presented here.

[1]Professors of soil chemistry, Department of Agronomy, University of Illinois, Urbana.

GROWTH CURVES

What is plant growth? Growth has been defined as an irreversible increase in the weight of a living organism. But such a definition may be made more meaningful if we consider growth in plants under field conditions as the process of fixation of solar energy. Such a concept would imply that plants, under field conditions, "grow" only when they are fixing solar energy. This, of course, is not strictly true, but for practical considerations growth and solar energy fixation go hand in hand.

In a consideration of plant growth as the fixation of solar energy, it may be helpful to review some relationships that are believed to be valid. First, one assumes that the plant has no voluntary choice in the distribution of the fixed energy among its various plant parts. That is, the genetic make-up of the plant predetermines the distribution of the fixed energy among the various plant parts for any given stage of development. Stated another way, the plant cannot say to itself, "I'm growing in an infertile soil, so I will have to grow a big root system to get the nutrients I need for growth," and then go ahead and do so. Second, one assumes that the genetic make-up of the plant determines the ratio of one plant part to another. That is, for any given plant species a definite ratio of roots to foliage to seed relationship exists, and this relationship is not affected by variations in the soil levels of an essential nutrient—for instance, between a deficient and a sufficient level. Actually, this assumption is implied by all who only use the weight of grain produced under a given field experimental condition, as the sole criterion for determining the influence of an environmental factor on the growth of a crop. For example, a study comparing root impedance of soils of varying degrees of compactness, where grain yields are the only measured crop criterion, assumes that the restricted root growth, if it occurs, will be reflected directly in the grain yield. While such an assumption, whether recognized or not, is usually valid, experimenters must always recognize that environmental factors, such as insects, diseases, climate, mobile nutrient deficiencies, etc., can disrupt, through late season damage, this ratio of grain to the rest of the plant much more easily than the ratio between roots and foliage. Whenever possible, total above ground growth should be measured in preference to grain yields alone.

If we think of growth as a process of solar energy fixation, it becomes evident that the amount of energy fixed (growth) is a function of some average rate at which energy fixation occurs and the time interval that the rate is in operation. Growth, G, may then be expressed as being equal to an average rate factor, R, times time, t, or $G = R \times t$, recognizing that the rate itself does change with different stages of plant development. The accumulation of plant mass as a function of time has been determined many times and always produces a typical sigmoid-shaped growth curve, as shown in Figure 1. Such a growth curve, in terms of daily increases in weight, shows that growth starts slowly, reaches a maximum (the grand period of growth), and then decreases as the plant matures. It can be

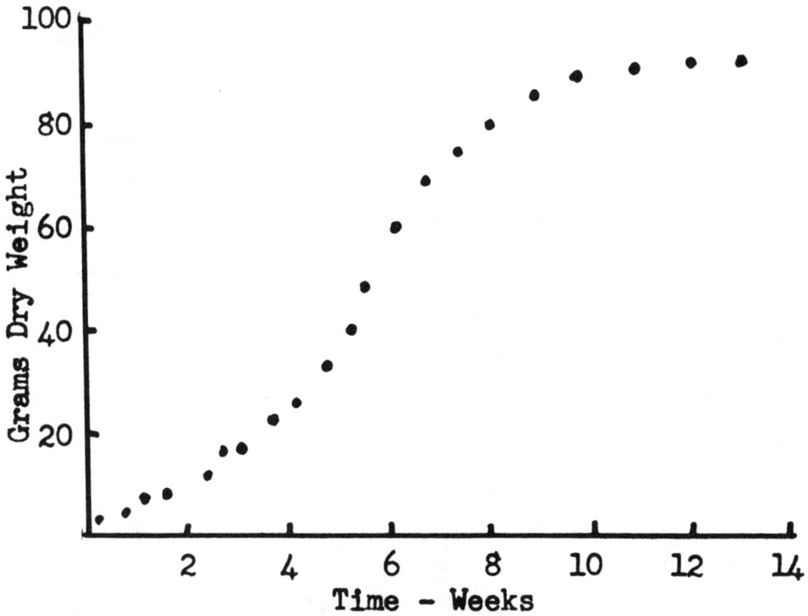

Figure 1. Sigmoid-shaped growth curve.

shown mathematically that such a growth curve is exponential in character, the sign of the exponent changing from plus to minus at the point where the plant has achieved about one-half of its total growth. Robertson (1907), expressed it logarithmically as follows: $\log\left(\frac{x}{A-x}\right) = k(t_o - t)$, where

k is a constant, A is the maximum dry weight, x is the dry weight at any time t, and t_o the time at which the weight of the plant is one-half of its final dry weight, the rate factor being whatever was dictated by the environmental factors under which the plants were grown.

In a fertility experiment, the experimenter is trying to determine the rate at which plants grow, and by carrying the experiment on for a sufficient length of time hopes that differences in rates will be reflected in the harvested weights at the end of some time interval, usually finite in terms of an annual crop. Changes in rate of growth will not change the shape of a growth curve, only the magnitude. From Figure 1 it is possible to calculate the constant, k, and the time, t_0 for the Robertson equation. By assuming values for A, one may calculate a family of curves (Fig. 2) showing the influence of different rates on the growth at any time, t. R-10 is the maximum rate; R-1 and R-2 represent rates that will give yields equal to 50 and 75% of the final R-10 yield. An interesting feature is that at any time t the various curves always bear the same relationship to each other. Therefore, if growth is related to time in an exponential fashion when rate is held constant, then growth must also be related to rate in an exponential fashion when time is held constant.

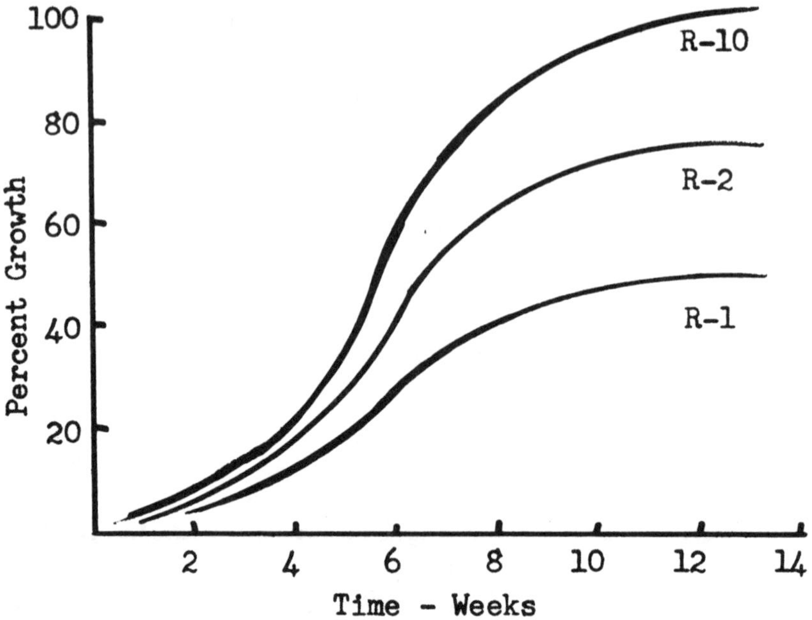

Figure 2. Sigmoid-shaped growth curves calculated for different growth rates.

The rate at which a plant grows is obviously not constant throughout the growing season. Growth starts out slowly, reaches a maximum rate, and then decreases as the plant goes into reproduction and senescence. Therefore, an environmental factor, like a nutrient deficiency, that influences the fixation of solar energy should be reflected proportionately over the time interval at all actual rates of energy fixation as dictated by the plant's various stages of growth. For example, a plant stunted by a potassium deficiency during formative stages of growth cannot compensate for its small leaf area during grain formation, even though potassium is added and the photosynthetic efficiency of the leaf area may be increased slightly. Since the change from vegetative growth to seed formation and senescence is controlled by the plant's auxin systems, and these in turn are controlled by photoperiodism, any decrease in plant development during an earlier period of growth cannot be overcome at later stages. The plant, like man, can never regain time lost.

Perhaps the most realistic way of deriving a growth curve, to express rate when time is constant, for different levels of an essential immobile nutrient in the soil, is the approach used by Mitscherlich and modified by Bray. This expression evolved gradually through the research of many workers, Mitscherlich (1913), Spillman and Lang (1924), Wilcox (1937), Bray (1944, 1954), White and Black (1954), van der Paauw (1952, 1962), Balba and Bray (1956) and the National Academy of Science (1961), and may well undergo further change. With slight modification the approach is as follows: It is assumed that every plant species has a finite growth possibility, A, when all nutrients and growth factors are adequate but not in harmful excess. Then if y represents a yield when a nutrient, b, is something less than adequate, one can say that for every level of b there must be a corresponding yield, y, or $dy = db$. But y can increase only to where $y = A$, so a limit is set on the possible changes in b within those values which are required to produce values of y which are always less than A. Therefore, $\frac{dy}{db} = c_1 (A - y)$, and $\frac{dy}{(A - y)} = db\, c_1$, which can be shown to become $\log (A - y) = \log A - c_1 b$, the Bray modified form of the Mitscherlich equation.

While Mitscherlich defined the value A as the ultimate yield when all growth factors were optimum, Bray has defined A as the yield possibility when all nutrients are present in adequate quantities, but not in harmful excess. But, for practical considerations, A may be taken as the yield obtained when any single nutrient under consideration is adequate, but not in harmful excess, provided all other immobile nutrients or growth factors remain unchanged and nitrogen is adequate for the attainment of the yield A.

Again, considering growth, under field conditions, as an expression of fixed energy, and recognizing that time is constant and finite for annual crops where grain yields are desired (that is, the grain is harvested at some time after maturity and death of the plant), then growth is always influenced by the rate of energy fixation and, conversely, any change in the rate must always affect growth in the same way, regardless of the cause for the change in rate, if the growth factor is continuous throughout the growth period. Or stated in another way, any given factor that decreases the growth rate over the growth period must be reflected in yield in the same way as any other factor that reduces the rate in the same amount. Therefore, if a rate reduction occurs, whether caused by variations in levels of an essential nutrient, weeds, sunlight, moisture, or soil compaction, the relationship between causative agents and growth should remain unchanged and subject to the same mathematical treatments.

APPLICATIONS OF THE MITSCHERLICH-BRAY EQUATION TO VARIOUS GROWTH FACTORS

The growth of a plant may be influenced by many factors. To get measurable reduced growth the rate of energy fixation must have been reduced over a significant period of time. Therefore, it should make little difference what growth factor reduced the energy fixation rate, the influence on plant growth should be the same. To illustrate this concept three environmental factors that influence yield are considered and evaluated in terms of an exponential growth function.

Weeds

It is well known that crop yields are reduced as weed intensity increases. Just how weeds bring about a reduction in the fixation of energy by a crop can vary with conditions. The relationship between weed intensity and yield decrease should be exponential in character. Two studies by Knake and Slife (1962) and Moolani, Knake and Slife (1964) may be used to illustrate the concept. The relationship between weed density and yield may be quantitized by using an exponential expression in the form of $y = A - c \log (1 + W)$, where y is the observed yield, A is the yield where no weeds were present, W is the number of weeds per 100 cm of row length and c is a constant. In practical terms it is more useful to use the percent concept where A is the 100% yield and y is the percent of A yield obtained for each level of W, the weed density but the constant c would be different. The original data and calculated values are given in Table 1. The 40 weeds/100 cm row length, since this is common to both experiments, was used to calculate the constant c for the equation given above, and if the exponential concept holds and the yields for the 40 weeds/100 cm are valid it is possible to calculate the percent reduction in yields for all the other weed densities and crops. These calculated values for both yield and percent are presented in Table 1. The excellent agreement between the found and calculated results indicate that the exponential function depicts accurately the relationship between the weed density and yield losses. A more accurate c may be calculated by using all of the experimental data and arriving at an average c for weed density in the crop. The predicted percent

Table 1. Corn and soybean yields as affected by pigweed and foxtail (data recalculated from Knake & Slife, 1962; Moolani et al., 1964)

Weeds/100 cm of row	Corn yields				Soybean yields			
	Found		Calculated		Found		Calculated	
	q/ha	%	q/ha	%	q/ha	%	q/ha	%
				Pigweed				
Band	38.2	61.6			11.2	44.6		
40	39.7	63.5	39.7	63.6	12.3	49.0	12.2	48.8
8	45.5	72.8	48.9	78.2	13.8	55.0	17.5	69.8
4	53.0	84.8	52.6	84.2	17.1	68.1	19.5	77.8
2	55.1	88.2	55.7	89.2	18.4	73.3	21.3	84.9
1	59.5	95.5	59.5	93.2	20.7	82.5	22.7	90.5
0	62.5	100.0	62.5	100.0	25.1	100.0	25.1	100.0
				Foxtail				
Band	44.9	75.5			18.8	71.7		
40	49.8	83.8	49.8	83.8	21.7	82.8	21.7	82.6
20	52.2	87.9	51.6	86.8	23.6	90.0	22.5	85.7
10	54.0	90.9	53.2	89.6	24.7	94.3	23.3	88.8
3.3	55.0	92.6	55.7	93.7	25.1	95.8	24.4	93.2
1.7	57.5	96.8	56.8	95.7	25.3	96.6	25.0	95.3
0	59.4	100.0	59.4	100.0	26.2	100.0	26.2	100.0

Table 2. Calculated percent yield reductions in corn and soybean yields from increasing densities of pigweed and giant foxtail

Weeds/100 cm of row	Corn		Soybeans	
	Pigweed	Foxtail	Pigweed	Foxtail
	%			
0	0	0	0	0
1	7	3	10	2
3	14	5	19	4
5	18	7	25	6
7	21	8	29	7
10	25	10	33	8
15	28	11	39	9
20	32	12	43	10
30	35	14	48	12
60	42	16	57	13
100	48	28	64	15

yield reduction for weeds may be determined by using the expression $y = A - c \log (1 + W)$ where y is the percent of growth at any weed level W, $A = 100\%$ growth, c is a constant and W is the number of weeds per 100 cm of row. Then by subtracting the percent growth from 100, the actual percent yield decreases for weeds are determined as shown in Table 2. The following c-values, 9.1 for the foxtail (*setaria faberii*)-corn (*Zea mays* L.), 7.4 for the foxtail-soybean (*Glycine max* L. *Merr*.), 23.7 for the pigweed (*Amaranthus* sp.)-corn and 32.2 for the pigweed-soybean systems were used in constructing the data in Table 2. These calculations extend the interpretative capability of the research beyond the usual statement that yield differences for the different weed densities were statistically significant.

Sunlight

Sunlight and rainfall are two components of the season that are known to influence yields. Shading can be shown to decrease plant growth in an exponential manner. Like many other growth factors sunlight influences growth in an exponential way. Working in the direction of increasing sunlight from 90% shade to full natural sunlight yields increase with increasing sunlight. Data by Earley (unpublished data, Illinois mimeo) on the influence of shading on corn yields has been recalculated in terms of sunlight and is presented in Table 3, which illustrates the exponential relationship between growth and sunlight. Here the percent reduction for each degree of shading has been calculated. The 'WF9 × C103' corn variety is bigger than 'Hy2 × Oh40.' Therefore, the actual weight lost by the WF9 × C103 is always greater than that lost by Hy2 × Oh40 but the percent loss from maximum weight is the same for both varieties. For both varieties the relationship between shading and percent sunlight is exponential and may be expressed by the equation $\log (A - y) = \log A - cb$, where

Table 3. Effect of shading on the total growth of two corn hybrids (E. B. Earley, University of Illinois, unpublished data)

Percent natural sunlight	Hy × Oh41 Growth	Hy × Oh41 Percent		WF9 × C103 Growth	WF9 × C103 Percent
%	g	%	1955*	g	%
100	563	100		733	100
70	535	95		632	86
40	441	78		549	75
30	363	64		317	43
20	319	57		299	41
10	130	23		155	21
			1956†		
100	591	100		703	100
70	473	80		597	85
40	377	64		484	69
30	275	47		325	46
20	170	29		170	24
10	49	8		58	8

* Shading started 8 July 1955.
† Shading started 10 June 1956.

A = 100%, y is the percent yield at any sunlight level b, and c is a constant. As the shading increases or percent sunlight decreases the proportionate decrease in corn yields increases. Based on the Mitscherlich-Bray function it is possible to calculate the yield decreases to expect from each percent increase in shading.

Moisture

A characteristic of the exponential function as modified by Mitscherlich-Bray is that when one growth factor is held constant it will exercise a constant proportional affect, or a constant percentage affect, on any other growth factor that is being varied. The data in Table 4, recalculated from Stanberry et al. (1955) illustrate the point where water and phosphorus were being varied. Note that for any given level of P the percent yield for the various water levels was approximately the same. For example, where 21.4 kg/ha of P was applied the percent yield for the various water regimes wet, medium, and dry were 74, 71 and 71% respectively. Similarly, for any given level of moisture the effect of moisture was the same for all rates of P applications. Thus, the medium water level always produced about 85% of maximum although the P treatments increased yields from about 64.3 to 91.0 metric tons/ha. These data indicate that water influences growth in an exponential way and that like P it is acting as an independent variable. If quantitative values for the water levels were available a table or graph could be constructed showing the percent yield losses for increasing moisture tensions.

Table 4. Effect of water and phosphorus on alfalfa yields (data recalculated from Stanberry et al., 1955)

	Water treatments					
	Dry		Medium		Wet	
P applied	Yield*	Percent	Yield	Percent	Yield	Percent
kg/ha	metric ton/ha	%	metric ton/ha	%	metric ton/ha	%
21.4	57.8	74	64.3	71	74.6	71
42.9	64.1	82	68.8	76	84.5	89
85.7	77.3	99	84.5	93	100.4	95
128.6	78.2	100	91.0	100	105.8	100

	P Treatment kg/ha							
Water regime	21.4		42.9		85.7		128.6	
	Yield	Percent	Yield	Percent	Yield	Percent	Yield	Percent
	metric ton/ha	%	metric ton/ha	%	metric ton/ha	%	metric ton/ha	%
Dry	57.8	77	64.1	76	77.3	77	78.2	74
Medium	64.3	86	68.8	81	84.5	84	91.0	86
Wet	74.6	100	84.5	100	100.4	100	105.8	100

* Yields are 4-year totals.

COMPARISON OF THE EXPONENTIAL AND QUADRATIC FUNCTIONS

While the biological base for preferring the exponential function over the quadratic function has been indicated a brief review of their similarities and differences may be of interest. In general terms the exponential function never reaches an absolute maximum. When one graphs, for example, soil test values against yield no matter how much of the nutrient is present in the soil the indicated yield never reaches a maximum or 100%. Similarly, the exponential will never indicate a toxic level or a decrease in yield from an excess of a nutrient or growth factor. By contrast the quadratic function reaches a maximum yield, e.g. 100%, with increasing levels of a nutrient or growth factor and then indicates a decrease in yield, or toxic levels, with further increasing levels of the growth factor. However, the quadratic function will indicate a toxic level or decrease in yield long before such toxicities occur under field conditions. In Figure 3 both functions indicate somewhat similar relationships between soil levels of K and percent yields during the first part of the function. The quadratic shows a steeper rise and reaches a maximum at about 310 kg/ha and then shows a rapid yield drop to almost zero growth at 558 kg/ha. The exponential function reaches a practical maximum of 99% at about 434 kg/ha and will never indicate a toxicity. Both expressions can give practical estimates of the nutrient level-plant growth relationship in the nutrient deficiency range, but both fail in predicting toxicity levels. It should be

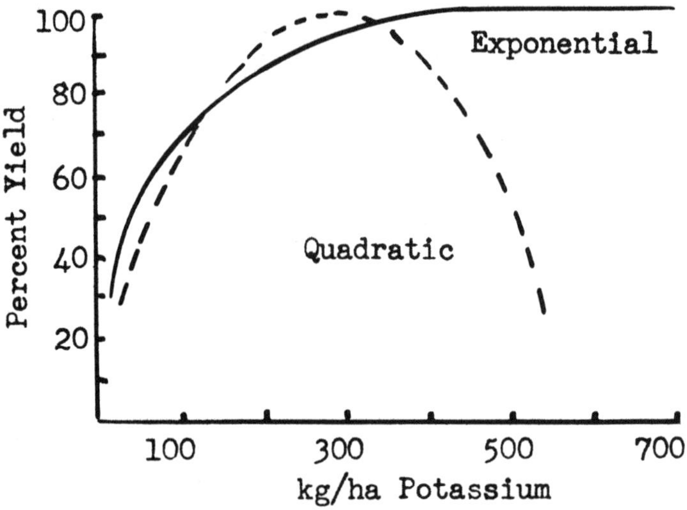

Figure 3. A comparative relationship between an exponential and a quadratic function.

noted that for any narrow upper range for both curves, say 85 to 95% sufficiency, where the experimental errors can become large, a straight line relationship may be found between yield and the levels of a growth factor. This often occurs when fairly high nutrient level soils are used for soil test correlation studies. However, the straight line obtained from such data permits no extrapolation beyond the narrow limits of the experiment.

In soil test correlation work where both soil nutrient levels as measured by soil tests and added fertilizer rates are being quantitatively related to growth response it becomes important that the mathematical function used represents the biological system over a wide range of nutrient levels. The objective of soil testing is to be able to predict a nutrient response independently of climatic, soil productivity, and management variables. Examples of how water, weeds, and seasonal variables may be related to crop response through the Mitscherlich-Bray growth function have been presented. If the relationship between the growth factor and growth is exponential in character, that factor is operating essentially as an independent variable. Water under normal field conditions and N limit yields in an absolute way, e.g. growth occurs until the supply is exhausted.

Water is peculiar in the sense that it influences growth exponentially at soil water tensions lower than the wilting point, and then when its supply to the plant is essentially exhausted it influences growth in an absolute way. Potassium and all other immobile nutrients cannot be completely exhausted, like water can, during a growing season. Under field conditions, except irrigation, the soil water tension is continually changing between rainfall renewals so that the soil's percent sufficiency with respect to water fluctuates from day to day. At low tensions the effect of water on growth is mild while at high tensions approaching the wilting point its effect on growth becomes severe. In the laboratory various soil water tensions can be established where the tension is kept constant but the water supply is unlimited. Under such conditions the various water stress levels can be measured and shown to affect yields in an exponential manner. But under field conditions, the water tension increases continually as the plant uses the water, thus always changing the growth rate slightly as the water tensions oscillate between zero and the 15 bar wilting point. Therefore, water's affect on yield, because of its changing intensity and length of stress, cannot be quantitatively measured between rainfall events, except where the supply is exhausted and plants wilt. The water stresses simply become, under normal field conditions, a contributing factor to the suitability of that season for crop production.

CORRELATING SOIL NUTRIENT LEVELS AND PLANT GROWTH

Agronomists have always been interested in relating plant growth and response to nutrient levels in the soil. The foregoing discussion pointed out that, in general, growth was a function of a rate of energy fixation acting through a given time period. In relating a level of a soil nutrient to growth, as is done in soil test correlation studies, one must assume that any given soil level is related to a particular rate of energy fixation; otherwise the growth-product of rate times time would not be expected to change. Actually, two assumptions are made: (i) that the nutrient acts essentially as an independent variable—that is, that a reduction in the quantity of the nutrient reduces the energy fixation rate independently of the other nutrient levels; and (ii) that some measured quantity of a soil nutrient level will, or can, be made to express, or define, the rate factor for that particular nutrient.

To understand the influence of a given level of a soil nutrient on the growth rate, it is necessary to classify soil nutrients according to their behavior in the soil. Usually only two broad classifications are needed: (i) those nutrients that are relatively immobile and (ii) those that are mobile in the soil. A relatively immobile nutrient is one that is adsorbed on the soil colloid, like the exchangeable cations, and therefore moves or diffuses slowly in the soil and is resistant to leaching. A mobile nutrient is one that moves freely in the soil, is not attracted to or held on the colloid surfaces, and moves both in and with the soil solution. This distinction, on the basis of mobility of soil nutrients, is fundamental to an understanding of their influence on growth.

A root may be divided into three general zones: (i) the growing tip and region of cell division and elongation, (ii) the zone of cell maturation and the root hair region, and (iii) the developed root with a mature epidermis and cortex layer. The zone capable of the most rapid nutrient absorption is the root hair region. Per unit of length, the root hair region is 10 to 25 times more efficient in absorbing nutrients than either of the other two zones. Therefore, as a root grows through a soil, the root hairs come into contact with the soil particles, or their absorbed water layers, extracting the immobile nutrients from the clay surfaces. But the root hair zone is continually moving forward as the root grows, and in effect is always feeding in "virgin" areas as the root explores the soil volume. The more mature root, with its much less efficient absorbing surface area, develops in the cylindrical volume of soil through which the root hairs have passed and extracted many, if not all, of the immobile nutrients.

A root system can never extract all of an immobile nutrient from a soil, as absorption is restricted to the surface area of the root system that can come into close proximity with soil particles. An immobile soil nutrient is therefore one whose diffusion rate is much slower than the penetration rate of the root hair zone into the soil. As a result, the renewal of soil nutrients in the area where the root hairs have passed is so slow, and the absorption efficiency of the mature root is so low, that the amount of renewal is essentially of no significance to the plant's growth. Under such conditions the roots are always feeding at a fairly uniformly low level if the soil is deficient in nutrients. Therefore, an immobile nutrient deficiency exerts a continuous and somewhat uniform influence on the average growth rate and should be reflected in growth in an exponential way.

A relatively mobile nutrient is one whose movement in the soil is more rapid than root elongation and movement of the root hairs through the soil, so that significant renewal of the nutrient in the feeding area is possible during the life span of the root hair. As a result, a mobile nutrient can be completely exhausted from the soil volume by the root system. Under field conditions, a plant is always growing at its maximum rate with respect to a mobile nutrient, such as nitrate nitrogen, as long as the supply lasts. When the roots have exhausted the supply, growth or the rate of energy fixation drops to near zero and for all practical purposes one can say that growth ceases. For this reason a mobile nutrient influences growth in an absolute way; that is, the growth is determined by the quantity of a mobile nutrient and the chemical composition of the crop.

Another factor that must be kept constant when immobile nutrients are being correlated with yield (growth) is plant population, or stand. Since differences in rate of growth are being determined as influenced by the nutrient level in the soil, it becomes mandatory that the stand, or the number of the plants involved in the final weight, be the same. A few more corn plants in one plot than in another, when extrapolated to hectare yields, can have a much greater effect on the weighted results than any differences in the rate of growth that may have resulted from treatment. This raises the question, when does a soil fertility experiment begin, with the introduction of the seed into the ground or when the seedlings are established to where the plants are independent of the seed? For soil test correlation studies the experiment starts when the plants are independent of their seed and all plot stands are equalized. Therefore, such factors as seed germination, shoot emergence and the survival of seedling plants to the point where they begin to fix their own energy must never be allowed to enter into the rate-of-growth study. This is not to say that studies of the effect of soil nutrient levels or other growth factors on seed germination, shoot emergence, and seedling survival are not valid and necessary, but, rather, that any effect a growth factor may have on germination and seedling survival must be separated from its effect on the plants subsequent growth rate.

For a crop like corn any constant plant population may be used, from a single plant to about 64,000 plants/ha, when studying the effect of an immobile nutrient on growth and yield. In a practical sense one objective in fertility research is to maximize energy fixation, or yield, per unit area, usually the hectare. Research involving plant densities, row widths or seeding rates are studies in maximizing energy fixation per unit area. An example of such a study is that of Welch, (1968), recalculated and presented in Table 5. Note that for all plant populations the percent sufficiencies for any given level of P are about the same, namely about 71% for the untreated and 92% for the 20 kg/ha rate. As population was increased the yields per hectare increased, the increase is exponential in character and follows the Mitscherlich-Bray growth function. From this relationship it is possible to calculate the appropriate constants and construct the data presented in Table 6, showing the influence of corn populations on yields per hectare.

Table 5. Effect of phosphorus and plant populations on corn yield

	Phosphorus treatment (kg/ha P)					
Plants/ha	0		20		40	
	quintal/ha	%	quintal/ha	%	quintal/ha	%
19,800	51.7	77	63.1	94	67.1	100
34,600	63.6	72	--	--	88.3	100
49,400	65.3	68	91.3	95	96.1	100
64,000	66.5	67	90.3	91	99.3	100
79,000	67.7	70	84.1	87	96.7	100

Table 6. Calculated relationship between corn population densities and percent of maximum yield

Population	Percent of Yield
plants/ha	%
64,200	100
49,400	98
44,500	97
39,500	95
34,600	93
29,700	90
24,700	85
19,800	78
14,800	70
9,900	53

In relating levels of an immobile nutrient to growth the Mitscherlich-Bray equation in the form $\log (A - y) = \log A - c_1 b$ is often used. This expression is sometimes referred to as the "percent yield" equation as it makes no difference whether absolute yields of A and y, or their percent relationship when $A = 100\%$ and y is given as a percent of A, are used in the above log equation. For example, if $A = 80$ quintals and $y = 40$ quintals, or if $A = 100\%$ and $y = 50\%$, the $c_1 b$ product in the equation remains unchanged. This means that a given level of an immobile soil nutrient is always capable of producing a definite percent of yield, and that it represents a given percent sufficiency. While many data have been accumulated to show the validity of this concept, the data in Table 7 are offered to illustrate the point. Of course, a decrease in the percent sufficiency for the soil is indicated between the years 1923 and 1973 because the level of K in the untreated RLP plots has declined through cropping during this period. During any short-time interval, such as 1936 to 1939, the percent sufficiencies remained quite constant. Note, too, that seasons varied the yields by more than 100% without affecting the percent relationship between the K treated and untreated plots. During this period the soil test values on the untreated plots dropped from around 80 kg/ha in 1923 to less than 45 kg/ha in 1973.

Table 8 presents data from the Brownstown field to illustrate the independence of one immobile nutrient from another. Note that the percent sufficiency of K was independent of the P treatment, and the percent sufficiency of P was independent of the K treatment for all three crops. These data tend to confirm that, within the range of nutrient levels commonly found in Illinois soils, a given level of an immobile soil nutrient is always adequate for a definite percent of yield over a wide range of seasons, nutrient levels, and soil conditions. Varying soil levels of an immobile nutrient are usually best correlated with growth by using the Mitscherlich-Bray function. The Mitscherlich-Bray function is not a

Table 7. Toledo field—corn yields and percent sufficiency of soil potassium for different years

Year	Treatments*		Sufficiency
	RLP	RLPK	
	quintal/ha		%
1923	19.2	32.4	59
1936	14.9	29.4	51
1937	22.3	41.7	53
1938	20.8	41.9	49
1939	27.6	54.7	51
1945	7.4	16.1	46
1955	24.0	56.7	42
1962	26.4	68.1	39
1971-1973	23.2	68.6	34

* R = Residues, L = Limestone, P = Rock Phosphate, K = Potash.

Table 8. Brownstown Field—crop yields and percent sufficiencies of potassium and phosphorus as independent variables

Treatment*	Corn Yields	Corn Percent sufficiency	Soybeans Yields	Soybeans Percent sufficiency	Wheat Yields	Wheat Percent sufficiency
	quintal/ha	%	quintal/ha	%	quintal/ha	%
			1945-1948 period			
			Potassium			
L	27.0		12.1		8.7	
LK_2	32.0	84	14.4	84	9.9	88
LN_2P_2	27.8		12.1		17.0	
$LN_2P_2K_2$	35.0	80	14.8	82	20.3	84
			Phosphorus			
L	27.0		12.1		8.7	
LP_2	35.0	89	12.3	98	13.1	66
LN_2K_2	34.1		13.6		11.2	
$LN_2K_2P_2$	35.0	97	14.8	92	20.3	55
			1971-1973 period			
			Potassium			
L	17.9		9.9		4.4	
LK_2	29.7	60	14.6	68	5.0	88
LN_2P_2	46.4		14.2		30.2	
$LN_2P_2K_2$	74.2	63	21.2	67	37.0	82
			Phosphorus			
L	17.9		9.9		4.4	
LP_2	20.8	86	12.3	81	8.2	54
LN_2K_2	63.7		17.0		19.7	
$LN_2K_2P_2$	74.2	86	21.2	80	37.0	53

* L = Limestone, N_2 = Highest nitrogen rate, P_2 = Highest phosphorus rate, K_2 = Highest potassium rate.

"growth curve" expression, since it assumes that time is constant. It will, however, express the accumulated growth obtained from different rates of a growth factor at any given constant time. It follows that if growth is to be related to different levels of immobile soil nutrients the resulting growth must be measured at some constant time, that is, by calendar days and not at a particular stage of development of the plant.

The data in Table 8 may be used to correlate soil test values with grain yields and for the determination of the constant c_1 in the Mitscherlich-Bray equation. For example, the K soil test was 80 kg/ha and the Bray P-1 test was 17 kg/ha on the Brownstown field in 1971. Applying these values in the Mitscherlich-Bray equation, and using the indicated percent sufficiencies for corn, we can calculate the c_1 values for K and P.

Potassium	Phosphorus
$A = 100, y = 61, b = 80$	$A = 100, y = 86, b = 17$
$\log(100 - 61) = \log 100 - 80c_1$	$\log(100 - 86) = \log 100 - 17c_1$
$1.5911 = 2.000 - 80c_1$	$1.1461 = 2.000 - 17c_1$
$80c_1 = 0.4089$	$17c_1 = 0.8539$
$c_1 = 0.0051$	$c_1 = 0.050$

By using several field locations with different K or P levels one can obtain an average c_1 value that may be applied to a wide range of soil types and climatic conditions. The following c_1 values are being used in Illinois with soil tests reported in kg/ha. These constants are the correlation constants in the Mitscherlich-Bray function and relate nutrient level to percent yield.

Constant	Corn	Soybeans	Wheat
c_1 for potassium	0.0048	0.0051	0.0065
c_1 for phosphorus	0.045	0.045	0.016

Once an average c_1 value has been obtained, arbitrary values of b may be selected that cover the full range of soil deficiencies and the corresponding y, or percent sufficiencies, determined. Similarly, arbitrary values of y may be selected and the corresponding value of b determined. Careful selection of values for y or b allow construction of a calibration curve, or table, relating soil nutrient levels to percent sufficiencies for use in interpreting the soil test values to the farmer.

The Mitscherlich-Bray function may be expanded to permit the calculation of the amount of fertilizer needed to raise the percent yield from any given starting level to any other desired upper level for which fertilization is desired. The expression then takes the form $\log(A - y) = \log A - c_1b - cx$, where c is the efficiency factor for the method of applying the fertilizer, and x is the quantity of fertilizer that needs to be added. The factor c is, of course, determined experimentally through fertilizer rate studies.

FACTORS THAT MAY INFLUENCE SOIL TEST CORRELATIONS

Good soil test correlations cannot be obtained unless the field yield data accurately measure the percent response to a given level of an immobile nutrient and the soil test accurately measures its available form. The Mitscherlich-Bray function has given good predictive results for both the Illinois K and P soil tests. If a good correlation is not obtained either the soil test is not extracting the proper form and amount of the nutrient present in the soil, the nutrient is not immobile or the yield data are not valid because of stand differences or insect and disease damage.

Crop Yields

Field experiments depending on growth or yield of grain as a criterion for differentiating between levels of immobile nutrients in soil are crude in terms of the accuracy with which the nutrient levels may be determined chemically. The best field experiments can seldom attain an accuracy of more than 5 to 7% and most probably carry an error of over 10%. It must be recognized, therefore, that differences in soil nutrient levels that can easily be measured chemically cannot always be detected by field techniques, especially if the response is less than 10%. As a result the most accurate soil test correlations are obtained by working with fairly deficient soils and at the lower end of the response curve where the response to the nutrient is large relative to experimental error and then extrapolating the data using the Mitscherlich-Bray function.

Adequate Mobile Nutrients

The harvested growth must represent the growth rate over the entire growing period. This can occur only when the supply of mobile nutrients, like nitrate-N, are sufficient to permit each growth rate to proceed to crop maturity or the designated harvest time. This may mean, in the case of N, that different N rates may be required for different rates of an immobile nutrient to avoid toxicity to plants at the slow rate of growth and to prevent N starvation of the plants at the higher growth rate.

Nutrient Mobility

For most soils that contain some inorganic and organic colloids and have a cation exchange capacity of over 3 meq/100 g of soil, probably only nitrate-N, B, and water can be considered mobile nutrients. However, as the colloid content and cation exchange capacity of soils decreases nutrients like K, Mg, Ca, and P become somewhat more mobile. In a quartz gravel culture, as used in many greenhouse studies, all of the nutrients are fully mobile. In the area between pure hydroponics and soils with exchange capacities over 3 meq/100 g there is a vast array of sands that cannot be fitted satisfactorily into a soil test correlation pattern. The difficulty lies in determining nutrient mobility changes that occur with decreasing exchange capacity. Therefore, correlation studies should not be conducted on sands as nutrient mobility determines the form of the growth function to be used. The Mitscherlich-Bray growth function applies only to the immobile nutrients. When a soil nutrient supply is partially immobile and partially mobile both the soil test extractant and the yield relationships fail. Sands are, therefore, undesirable areas for soil test correlation studies.

SUMMARY

Correlating levels of a soil nutrient with growth and response to fertilizers is not an easy task nor can it be done in a haphazard manner. Careful selection of the site cannot be overemphasized. The growth characteristics of the crop, the range of what constitutes adequate but not excess fertilization, the correct measurement of the available nutrient form, the experimental design needed to measure the growth response accurately, a basic knowledge of how nutrients behave in the soil and how plants absorb and respond to them are but a few of the "knowledge" prerequisites for good correlation studies. Careful attention to the crop's biological requirements, as well as statistical design, will lead to good soil test correlations and ultimately to an effective soil testing program.

LITERATURE CITED

Balba, A.M., and R. H. Bray. 1956. New fields for the application of the Mitscherlich type equation. 1. A quantitative measure for the relative effectiveness of the nutrients. Soil Sci. 82: 497-502.

Bray, R. H. 1944. Soil-plant relationships. 1. The quantitative relation of exchangeable potassium to crop yields and to crop responses to potash additions. Soil Sci. 58: 305-324.

Bray, R. H. 1954. A nutrient mobility concept of soil-plant relationships. Soil Sci. 78: 9-22.

Knake, E. L., and F. W. Slife. 1962. Competition of *Setaria faberii* with corn and soybeans. Weeds 10:26-29.

Mitscherlich, E. A. 1913. Bodenkunde fur Land und Forstwitre. Paul Parey, Berlag., Berlin.

Moolani, M. K., E. L. Knake, and F. W. Slife. 1964. Competition of smooth pigweed with corn and soybeans. Weeds 12:126-128.

National Academy of Sciences, National Research Council. 1961. Status and methods of research in economic and agronomic aspects of fertilizer response and use. Publication 918. National Academy of Science, Washington, D.C.

Paauw, F. van der. 1952. Critical remarks concerning the validity of the Mitscherlich effect law. Plant Soil 4:97-106.

Paauw, F. van der. 1962. Fertilization with phosphorus. International Superphosphate Manufacturers Association Spec. Bull. No. 32.

Robertson, T. B. 1907. On the normal rate of growth of an individual and its biochemical significance. Arch. Entwicklungsmech. Organismen 25:580-614.

Spillman, W. J. and E. Lang. 1924. The law of diminishing returns. World Book Company, Chicago, Ill.

Stanberry, C. O., C. D. Converse, H. R. Haise, and O. J. Kelley. 1955. Effect of moisture and phosphorus variables on alfalfa hay production on the Yuma Mesa. Soil Sci. Soc. Am. Proc. 19: 303-310.

Welch, L. F. 1968. Corn population versus P need. Illinois Fertilizer Conference Proceedings, Illinois Fertilizer Industry Association.

White, W. C., and C. A. Black. 1954. Wilcox's agrobiology: III. The inverse yield-nitrogen law. Agron. J. 46: 310-315.

Willcox, O. W. 1937. ABC of agrobiology. W. W. Norton and Company, Inc., New York.

Partitioning of Soil Test—Crop Response Probability

L. A. Nelson and R. L. Anderson[1]

Much of the research in the soil testing area has dealt with developing chemical procedures for evaluating the levels of available nutrients in the soil. Techniques must be available for relating results of soil tests using these procedures to crop yield response before such results are useful. Colwell (1967) emphasized the importance of this relationship when giving a definition for a soil test, "A measurement qualifies to be termed a soil test for a particular nutrient if, and only if, it provides information on the fertilizer requirement of a crop for that nutrient."

PREVIOUS MODELS

Some of the work which has been done in the correlation area was reviewed by Fitts and Nelson (1955). They discussed the use of the product-moment correlation coefficient, r, for relating percentage yield to soil test results and pointed out some of the limitations of this approach. They also described the "probability" approach as suggested by Fitts (1955). The probability of obtaining a profitable increase in yield for a given fertilizer treatment is plotted against soil test results. The probability is the percentage of fields within a given range of soil test values that respond profitably to application of fertilizer. Fitts reported that in the "probability" approach, the odds are against receiving a profitable increase from fertilization on fields testing in the high classes. Individual year conditions, however, could result in responses from these soils testing in high classes, or lack of responses from soils testing in low classes.

Bray (1948) following the earlier work of Mitscherlich, Baule, and Wilcox (Spillman & Lang, 1924) fitted percentage yield versus soil test values using a modified Mitscherlich growth curve (curve of the law of

---

[1]Professors, Departments of Statistics, North Carolina State University and University of Kentucky, respectively.

diminishing returns). He did extensive work with both P and K. The modified Mitscherlich equation which he developed was:

$$\log(100 - Y) = \log 100 - c_1 b_1 \qquad [1]$$

where 100 = percentage yield of crop with all nutrients at adequate level,
Y = percentage yield of crop with all nutrients except nutrient being studied at an adequate level,
c_1 = proportionality constant, and
b_1 = soil test value of the nutrient for the untreated plot (e.g., K).

The c_1 values calculated for the individual fields used in the correlation study were averaged and the resulting mean c_1 was used to construct a continuous curve. Methods of calculating an individual c_1 when $Y \geqslant 100$ [and $\log(100 - Y)$ is indeterminate] were not given. Altering Y data to permit taking logarithms of $(100 - Y)$ results in an estimation procedure for c_1 which is not least squares.

Colwell (1967) developed a model which involved obtaining orthogonal polynomial coefficients for the regression of yield responses on various powers of $\sqrt{X}$, where X is level of added nutrient. He then regressed on the soil test values the linear and quadratic coefficients, which had explained most of the within-location variability. He was able to estimate the optimal rates of fertilizer using estimates of parameters from his final model and cost/price information. His method involves heavy computations which may be programmed for a digital computer.

A number of workers have used some type of regression model which regresses actual yield on both added nutrient and soil test nutrient either as separate terms or combined in the same variable. Model specification appears to be a major problem here as discussed by Laird and Cady (1969). The influence which is attributed to soil test would depend to a large extent upon what other variables are included and how they are modeled. Also, the tests of significance may be of questionable validity due to heterogeneity of errors and correlations among the errors within site.

YIELD VARIABLES

A number of yield related or derived variables have been used for evaluating the relationship between yield and soil test results. Some of these are:

1) Y_0 = check yield with 0 level of nutrient being studied,
2) Percentage yield = $(Y_0/Y_{max}) \times 100$, where Y_{max} = yield with all factors at adequate but not excessive levels,
3) Percentage yield increase = $[(Y_{max} - Y_0)/Y_{max}] \times 100$,
4) Actual yield increases: $\Delta Y_1 = Y_1 - Y_0, \Delta Y_2 = Y_2 - Y_0 , \ldots ,$
$\Delta Y_{max} = Y_{max} - Y_0$.

Actual yields (e.g., Y_0) usually do not correlate well with soil test results because they contain variation among locations attributable to climatic variables and also to variation in soil texture, structure, CEC, source of charge, and organic matter content.

Percentage yield has been used as the response variable in most soil test correlation studies in the U.S. This variable correlates well with soil test levels only in the case of immobile nutrients such as P and K. Bray (1948), who used percentage yield extensively in correlation studies involving these nutrients, discussed the theory behind Baule's percentage yield concept.

Percentage yield has a serious agronomic limitation. It is difficult to interpret directly in terms of amounts of fertilizer to add or profitability of same. It also has certain statistical limitations owing to the fact that it is calculated using a different denominator for each location; hence, the variance may not be homogeneous for all the data. This could cause problems in analysis of variance or in testing hypotheses in connection with regression.

We prefer to use data based on actual yield increases (ΔY or ΔY_{max}) rather than percentage yield. Both of these variables are designed to remove from the data location to location variability due to factors other than the soil nutrient being studied. Actual yield increase has a definite advantage in that direct economic interpretations can be made from the results of soil test correlation studies.

Both percentage yield and actual yield increase are subject to influence by interaction effects between a nutrient and the location variables. Bray (1948), however, concluded that these location variables (primarily soil factors) do not affect the correlations as much as had been hypothecated and that the influence of all other soil properties on the availability of a given form of nutrient has been overestimated in comparison to the importance of the amount of nutrient present in the soil.

Anderson and Nelson (1975) described a plateau response pattern for certain soils, particularly in developing countries where limitations of certain factors force a ceiling on the yield. They suggested fitting a series of models belonging to the linear-plateau family and selecting a particular member of the family which provides the best fit. Prediction with the model chosen provides an estimate of the plateau yield. Often, several treatment levels may be on the plateau and in this case, the plateau average should be used in calculating ΔY_{max} or percentage yield. This is advantageous statistically because it will produce more precise estimates of either variable than if a single value had been used in estimating Y_{max}.

DATA FOR SOIL TEST CORRELATION

Regardless of model and response variable to be used, reliable data must be available for good soil test correlation work. Since the important agronomic aspects which help insure that the data are reliable have been discussed in the review article by Fitts and Nelson (1955), we will focus attention upon experimental design aspects.

Experimentation for Soil Test Correlation

Experiments are necessary to provide data for soil test correlation. Either multilevel fertilizer rate experiments or none versus adequate rates for the nutrient being studied are used. The former are probably more efficient because they provide estimates of optimal fertilizer rates in addition to soil test correlation information.

Some 20 site-years of data should be sufficient in most cases to provide a reliable estimate of the response and its correlation with soil test results.

As in all experimentation, precision is increased by controlling as many factors as possible except those being intentionally varied. If a large series of soil fertility experiments is conducted, the experimental design, cultural operations, etc., should be constant throughout the series. Bray (1948) discussed some of the experimental design aspects. For correlation work, he described a simple two-treatment experiment, with three or four replications per field. Using this experiment, one plot is treated so that all nutrients are adequate or in excess but not in harmful excess. An adjacent plot receives the same treatment except that the nutrient studied is omitted from the treatment.

Bray also suggested an alternative experimental design which involves several levels of the nutrient. It was recommended that the plot with no nutrient added should be located in the center with the plots receiving the various increments adjacent to the four sides and the four corners. We favor random allocation of treatments to plots to assure unbiased estimates of experimental error and to insure that a treatment will not be favored or disfavored by being assigned to any particular plot.

Anderson and Nelson (1975) described several experimental designs which may be useful in soil fertility response work. Complete factorials were recommended for experiments in which the number of nutrient factors is small (say 2). All designs recommended were replicated and randomized.

PARTITIONING SINGLE NUTRIENT RESPONSE DATA

Cate-Nelson Graphical Method

Cate and Nelson (1965) developed a graphical method for partitioning the two-dimensional percentage yield versus soil test level scatter into two groups. Using the method on graphs of percentage yield versus soil test level, it is possible to classify soils into two groups: (i) those for which probability of response to fertilizer is large and (ii) those for which probability of response to added fertilizer is small. The original concept of such a partition was developed for soils of certain countries of Latin America where limited soil fertility information was available. A classification of soils into these groups, although perhaps not as precise and detailed as might be desired, did provide some benchmarks upon which further research and interpretation might be based. Sites having soils with test values below the critical level could be selected for the bulk of future fertilizer response trials. It might also be possible to draw interpretive maps showing extensive areas in a developing country where available soil test results indicate that the soils are in the "low" group and should, therefore, respond to fertilization.

The soil test value separating the two groups described above was termed the "critical level" by the authors. This was a rather general use of the word "critical" and does not imply that this level of nutrient is absolutely essential for growth and development.

The Cate-Nelson graphical method is based upon dividing the *Y-X* scatter diagram into four quadrants and maximizing the number of points in the positive quadrants while minimizing the number in the negative quadrants. The original idea of using the distribution of points in the four quadrants came from a nonparametric test of association developed by Olmstead and Tukey (1947). Steps in the Cate-Nelson Graphical Method are as follows:

1. Percentage yield values are obtained for a wide range of locations where results of fertilizer rate studies are available. Some consideration should be given to what is the underlying population of farms which is being sampled.

$$\text{Percentage yield} = \frac{\begin{array}{c}\text{Yield at 0-level of nutrient being}\\ \text{studied; other factors at adequate}\\ \text{but not excessive levels}\end{array}}{\begin{array}{c}\text{Yield where all factors are at}\\ \text{adequate but not excessive levels}\end{array}} \times 100. \quad [2]$$

   The numerator will be the check mean for the nutrient being studied. For the denominator, a plateau average should be used in cases where a linear-plateau response pattern is obtained. Otherwise, maximum yield should be used.
2. Soil test values for the nutrient being studied should be obtained for the locations mentioned above. The check plot test values should be averaged. Thus, there will be a single percentage yield and one soil test value for each location.
3. The scatter diagram of percentage yield (*Y*-axis) versus soil test value (*X*-axis) is plotted on arithmetic paper. The range in values on the *Y*-axis is 0 to 100%. Occasionally percentage yields of greater than 100% are obtained due to random effects. They should be plotted as calculated.
4. A piece of clear plastic having roughly one and one-half the dimensions of the graph is cut out for use as an overlay. A pair of intersecting perpendicular lines is drawn on the overlay with black India ink in such a way that it is divided into four sectors having

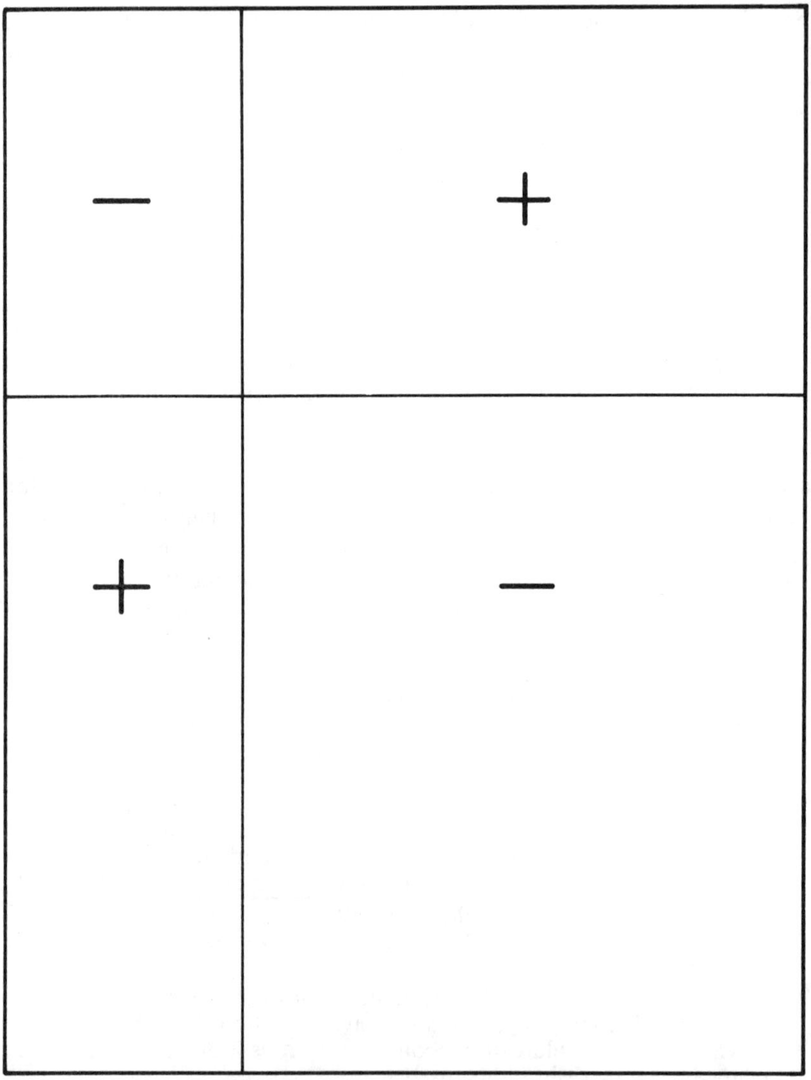

Figure 1. Format of clear plastic overlay used in Cate-Nelson graphical method. Cate & Nelson, 1965.

areas of roughly the same relative size as shown in Figure 1. The signs of these four quadrants are then labeled with black India ink.

5. The overlay is moved about horizontally and vertically on the graph, always with the two lines parallel to the two axes on the graph, until the number of points showing through the overlay in the two positive quadrants is at a maximum (or conversely, the number of points in the negative quadrants is at a minimum).
6. The positions of the lines on the overlay with respect to the axes of the graph are transferred to the graph by making marks on the

edges of the graph. The two intersecting lines are then drawn lightly on the graph with pencil. The point where the vertical line crosses the *X*-axis is defined as the "critical" soil test level.

An example of a Cate-Nelson scatter diagram of percentage yield versus soil test value for the Bray No. 1 Method of P extraction is shown in Figure 2. The data are from the Soil Test Work Group for the National Soil Research Committee (1956).

Cate-Nelson scatter diagrams of similar quality have been developed for P and K for several Latin American countries by the staff of the International Soil Fertility Evaluation and Improvement Program of North Carolina State University (1971). Examples are shown in Figure 3 for P fertilization of wheat (*Triticum aestivum* L.) and potatoes (*Solanum tuberosum*) in Bolivia.

In addition to its simplicity, the Cate-Nelson method

1) provides a definite decision point, thus forming the basis for a simple classification of soils on the basis of soil test results,
2) does not require elaborate computational equipment,
3) does not require the usual assumptions necessary in ordinary statistical procedures, and
4) results in partitions upon which individual extreme points have little influence.

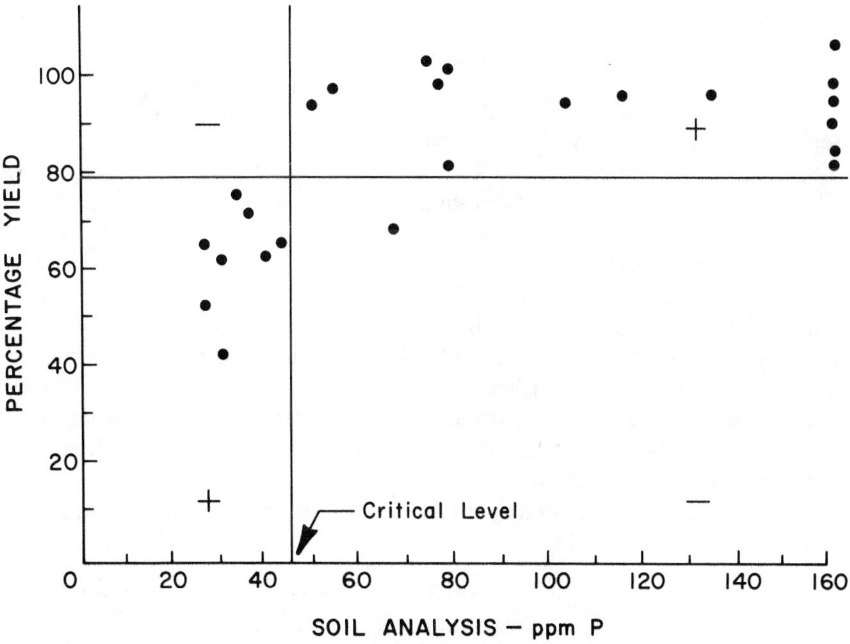

Figure 2. Scatter diagram of percentage yield of millet (*Setaria* sp.) versus soil test P by Bray #1 method for data of Soil Test Work Group. Cate & Nelson, 1971.

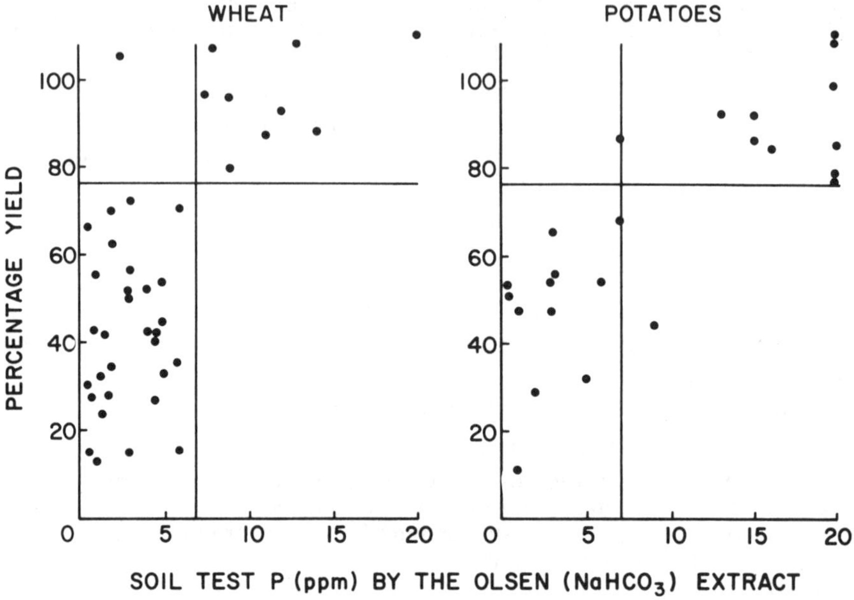

Figure 3. Scatter diagrams of percentage yield of wheat and potatoes versus soil test P by Olsen method for Bolivian data. International Soil Fertility Evaluation and Improvement Program, 1972.

An Analysis of Variance Method[2]

Cate and Nelson (1971) reported an iterative statistical procedure for separating percentage yield data into two or more classes based upon maximization of the class sum of squares in a one-way analysis of variance. This sum of squares reflects the weighted sum of squares of the differences between the percentage yield means for the various classes and the grand mean. In other words, using this procedure, one finds quantitatively the best divisions from the point of view of maximizing mean differences among classes. These, in turn, should be the best divisions from the point of view of prediction. The method was reported to form a more precise basis for defining fertilizer recommendation classes (e.g., "low," "medium," and "high"), which often have been defined rather subjectively and arbitrarily in the past. Choice of the number of classes, k, is discussed in another section of this paper.

The procedure described in their paper as applied to data to be divided into two classes is as follows:

1. Percentage yield (Y) is calculated for each location in the same

---

[2]This Cate-Nelson Analysis of Variance Method has been adapted by James Goodnight, Department of Statistics, North Carolina State University and SAS Institute, Inc., for computations using the Statistical Analysis System (SAS) of A. J. Barr and J. H. Goodnight. Department of Statistics, North Carolina State University, Raleigh. NIH Project No. FR-00011.

manner as used in the Bray or Cate-Nelson graphical method.

2. Soil test values, X, are obtained for each location.
3. The data are ordered in an array based upon rankings of the X values, i.e., soil test values. The (X,Y) pairs are maintained in this order throughout the analyses.
4. Starting with the partition that will place two points to the left and ending with that which will place two to the right of the partition, the class sum of squares is calculated according to techniques used in an analysis of variance for a one-way classification, i.e.,

$$\text{Class Sum of Squares} = \frac{\begin{bmatrix}\text{Total of percentage} \\ \text{yields for Class 1}\end{bmatrix}^2}{n_1} + \frac{\begin{bmatrix}\text{Total of percentage} \\ \text{yields for Class 2}\end{bmatrix}^2}{n_2} - \frac{[\text{Total of percentage yields for both classes}]^2}{n} \qquad [3]$$

where n_1 = number of observations in Class 1 (to the left of the partition);
n_2 = number of observations in Class 2 (to the right of the partition); and
n = total number of observations $(n_1 + n_2)$.

In successive steps the partition is moved one point to the right and a new Class Sum of Squares is calculated until the partition is two points from the right.

5. The Class Sum of Squares is then divided by the Total Corrected Sum of Squares for percentage yields to obtain R^2. Total Corrected Sum of Squares $= \Sigma y^2 = \Sigma Y^2 - (\Sigma Y)^2/n$.
6. By this simple iterative process, a series of R^2 values is obtained for divisions made at various levels of X. The critical level of X is taken as that level of X-division for which R^2 is maximum.

A constant divisor (Σy^2) is used in calculating all R^2 values. Hence, one could achieve the same results by maximizing the Class Sum of Squares. The R^2 was calculated by the authors in order to compare the fit using the procedure with that using other models such as the Bray, linear, quadratic, logarithmic, and reciprocal. See the next section for an example of the computations in Step 4, where ΔY_{max} is used for Y.

Although the Cate-Nelson Analysis of Variance Method (ANOVM) is generally superior to the Graphical Method (GM) in estimating optimal partitions, the two methods seem to produce quite similar results. Therefore, since ANOVM can produce biased results if there are outliers, it may be preferable to use GM for data with a number of suspected aberrant observations.

USE OF ACTUAL YIELD INCREASE WITH CATE-NELSON ANALYSIS OF VARIANCE METHODS

As discussed in the section on Yield Variables, actual yield increase (ΔY_{max}) has the advantage over percentage yield of permitting direct economic interpretations.

Table 1. Cotton yield data of Freitas et al. (1966) and soil test K together with results using two and three class models (Cate-Nelson Analysis of Variance Method)

Y_{max}	Y_0	ΔY^*_{max}	Soil test K	Partition between†	Class sum of squares 2-class model	Range in class sum of squares 3-class model
———	kg/ha	———	——	ppm ——		
1604.0	841.8	762.2	26			
1991.0	1287.8	703.2	28	28-30	482,656	558,098-1,020,848
1044.8	653.5	391.3	30	30-31	435,837	503,650- 951,555
1654.0	674.3	979.7	31	31-34	957,957	967,084-1,175,762‡
1124.0	908.5	215.5	34	34-35	764,944	792,195- 992,452
875.5	616.8	258.7	35	35-40	670,038	693,320- 924,895
1345.0	846.8	498.2	40	40-44	784,501	793,593- 922,122
1363.0	856.3	506.7	44	44-49	921,247	921,989- 982,144
1040.5	979.0	61.5	49	49-56	722,346	724,022- 819,346
3072.3	3041.8	30.5	56	56-58	545,522	549,684- 636,924
1892.8	1257.3	635.5	68	68-75	780,873	780,878- 824,487
1080.8	1112.8	−32.0	75	75-77	578,244	578,573- 629,783
1161.8	1131.0	30.8	77	77-78	446,509	446,520- 504,091
2820.8	2404.3	416.5	78	78	546,348	546,375- 584,318
2746.0	2779.3	−33.3	78	78-102	394,453	394,503- 442,193
2622.0	2543.0	79.0	102	102-118	319,905	319,905- 371,048
2329.0	2255.3	73.7	118	118	251,877	252,121- 308,427
1451.8	1423.0	28.8	118	118-132	173,647	184,638- 245,196
1370.3	1175.0	195.3	132	132-133	168,593	173,958- 232,423
2165.5	1997.8	167.7	133	133	156,494	215,251
3330.5	3220.0	110.5	133	133-153	124,403	
2509.5	2214.0	295.5	153	153-193	198,678	
1539.0	1651.0	−112.0	193			
1148.0	1114.8	33.2	211			

* The total sum of squares for ΔY_{max} is 1,931,734.

† This partition is between "low" and "high" in the two class model and between "low" and "medium" in the three class model.

‡ The numbers of observations in the "low," "medium," and "high" classes for the optimum partitioning (Class SS = 1,175,762) are 4, 7, and 13, respectively.

In this paper, we will discuss the use of the Analysis of Variance techniques for ΔY_{max} values. An example of the original Cate-Nelson Analysis of Variance procedure is shown in Table 1 for the ΔY_{max} data of Freitas et al. (1966). The total corrected sum of squares is 1,931,734 which was obtained using the formula given in Step 5 in the previous section. One possible partition is the first two observations versus the other 22. Using ΔY instead of percentage yields in Eq. [3], the corrected sum of squares for this partition is

$$\text{Class Sum of Squares} = \frac{[1465.4]^2}{2} + \frac{[4831.3]^2}{22} - \frac{[6296.7]^2}{24} = 482{,}656.$$

The Class Sums of Squares for various other two-class partitions are obtained in a similar manner, using sums and divisors which depend upon the partition. As shown in Table 1, the Class Sum of Squares is maximum when the partition is placed between 31 and 34 ppm of soil test K. The "critical level," $CL = (31 + 34)/2 = 32.5$ ppm of soil test K.

The group means of ΔY_{max} for the two groups partitioned at 32.5 ppm are 709.1 and 173.0 kg/ha. This would imply that applying adequate K to cotton (*Gossypium* sp.) according to soil test recommendations for soils testing "low" in K would result in an average response of 709.1 kg/ha as compared with an average response of 173.0 kg/ha for the "high K" class. One could assume prices for cotton and fertilizer and determine how much fertilizer would be profitable on the average for soils testing in each of the two classes.

The computational techniques described for the Cate-Nelson Analysis of Variance Method may be extended to three or more classes. The Class Sums of Squares are calculated for the various partitions involving classes having at least two observations. The formula for computing the Class Sum of Squares for three classes is

Class Sum of Squares (3 Classes) =

$$\frac{\left[\begin{matrix}\text{Total } \Delta Y, \\ \text{Class 1}\end{matrix}\right]^2}{n_1} + \frac{\left[\begin{matrix}\text{Total } \Delta Y, \\ \text{Class 2}\end{matrix}\right]^2}{n_2} + \frac{\left[\begin{matrix}\text{Total } \Delta Y, \\ \text{Class 3}\end{matrix}\right]^2}{n_3} - \frac{\left[\begin{matrix}\text{Total } \Delta Y, \\ \text{All Classes}\end{matrix}\right]^2}{n} \quad [4]$$

where n_1 = number of observations in Class 1 (low class);
n_2 = number of observations in Class 2 (middle class);
n_3 = number of observations in Class 3 (high class); and
n = total number of observations ($n_1 + n_2 + n_3$) .

The computational work and also the number of classes having only a few observations is increased considerably by an increase in the number of classes.

Results of partitioning the Freitas et al. (1966) data into three classes are given in Table 1. The optimum partitions for "low" to "medium" and "medium" to "high" are (31 + 34)/2 = 32.5 and (68 + 75)/2 = 71.5 ppm K, respectively. Average ΔY_{max} for the "low," "medium," and "high" classes are 709.1, 315.2, and 96.4 kg/ha cotton, respectively.

There are two major difficulties with the use of the Cate-Nelson Analysis of Variance method:

(1) there is no unique method of estimating the error variance and
(2) there is no acceptable method of testing for the optimal number of classes.

In addition, since the ΔY_{max} data are much more erratic than the percentage data, there may be difficulty in selecting the best division points. We recommend consideration of the modification presented below for this type of data.

MODIFIED ANALYSIS OF VARIANCE METHOD

Because of the erratic nature of the ΔY_{max} values, it is suggested that these data be combined into subgroups of at least two observations each, as indicated in Tables 1, 2, and 3. The following principles are used in constructing the boundaries of the subgroups. These are illustrated by the data from Table 1.

Table 2. Corn yield data of Bray (1944) ordered by soil test K together with sums of squares of ΔY_{max} for various partitions of two and three class models

Y_{max}	Y_0	ΔY_{max}*	Soil test K	Partition between†	Class sum of squares 2-class model	Range in class sum of squares 3-class model
—— quintal/ha	——	——	— ppm —			
34.7	13.3	21.4	25			
44.5	20.6	23.9	25	25-33	496	517-1061‡
28.9	17.0	11.9	33	33-36	453	470-999
35.3	17.5	17.7	36	36-38	601	611-1001
44.6	23.0	21.6	38	38-38	881	883-1046
37.8	25.1	12.8	38	38-43	923	923-993
34.0	21.2	12.8	43	43-45	985	985-1016
33.8	31.4	2.4	45	45-62	787	787-827
28.0	22.6	5.4	62	62-78	701	701-735
67.7	63.5	4.2	78	78-79	609	609-646
56.0	49.9	6.1	79	79-87	572	572-580
53.8	46.3	7.4	87	87-106	572	572-585
39.7	39.2	0.5	106	106-106	446	447-457
43.6	39.8	3.8	106	106-110	396	397-415
57.8	60.5	-2.8	110	110-111	249	249-253
63.1	61.0	2.1	111	111-112	192	192-193
51.6	50.7	0.9	112	112-118	127	127-129
62.5	56.3	6.2	118	118-120	131	133
59.0	57.2	1.8	120	120-130	86	
41.6	39.8	1.8	130	130-132	45	
62.3	59.0	3.3	132			
65.7	62.7	3.0	161			

* The total sum of squares for ΔY_{max} is 1230. All Y figures converted to metric, so occasional slight rounding discrepancies may occur.

† This partition is between "low" and "high" in the two class model and between "low" and "medium" in the three class model.

‡ The numbers of observations in the "low," "medium," and "high" classes for the optimum partitioning (Class SS = 1061) are 2, 5, and 15, respectively.

(a) There must be a substantial decrease in ΔY_{max} between the two values on each side of a division line, and the average ΔY_{max} should be decreasing. There is no division between 703.2 and 391.3, because the next observation jumps to 979.7; since there is a substantial decrease after 979.7, the division line is drawn between 979.7 and 215.5.

(b) The division should not separate two identical (or almost identical) values of Soil Test into two different subgroups, e.g., do not divide between ΔY_{max} of 416.5 and −33.3 because Soil Test K is 78 ppm for both. However, an increase in Soil Test K should not automatically produce a division; for example, there is no division between −33.3 and 79.0 because ΔY_{max} is increasing.

The purpose of this subgrouping is three-fold:

(1) Smooth the values of ΔY_{max} used in subsequent analyses.
(2) Use the within subgroup variance as an estimate of the error variance. One can also test for variance heterogeneity if desired.
(3) Provide a first step towards selecting the best division points.

Table 3. Tomato yield data of Arnold and Schmidt (1951) ordered by soil test P together with sums of squares of ΔY_{max} for various partitions of two and three class models

Y_{max}	Y_0	ΔY_{max}*	Soil test P	Partition between†	Class sum of squares 2-class model	Range in class sum of squares 3-class model
metric tons/ha			ppm			
28.0	8.1	19.9	10			
28.7	12.5	16.1	12	12-13	241	301-532
18.8	9.0	9.9	13	13-14	211	266-492
28.4	15.5	13.0	14	14-18	251	326-487
27.6	18.1	9.4	18	18-27	241	306-467
40.1	22.4	17.7	27	27-28	371	417-512
13.2	8.3	4.9	28	28-29	296	356-462
30.2	15.7	14.6	29	29-30	387	417-492
35.2	28.4	6.7	30	30-34	351	376-452
43.0	28.2	14.8	34	34-38	457	462-537‡
37.4	31.6	5.8	38	38-43	417	422-492
35.2	29.8	5.4	43	43-48	382	382-452
34.3	31.6	2.7	48	48-52	311	331-387
28.0	24.4	3.6	52	52-54	261	281-356
59.4	48.8	10.5	54	54-55	311	321-371
62.7	53.3	9.4	55	55-58	351	356-402
36.7	34.5	2.2	58	58-64	291	296-341
34.5	30.9	3.6	64	64-64	256	261-306
34.5	30.9	3.6	64	64-69	226	231-276
34.3	29.3	4.9	69	69-70	216	241-261
9.4	9.6	-0.2	70	70-71	141	201
51.7	43.9	7.8	71	71-72	186	
22.0	18.4	3.6	72	72-85	181	
47.5	52.0	-4.5	85			
40.8	39.6	1.1	132			

* The total sum of squares for ΔY_{max} is 865. All Y figures converted to metric, so occasional slight rounding discrepancies may occur.

† This partition is between "low" and "high" in the two class model and between "low" and "medium" in the three class model.

‡ The numbers of observations in the "low," "medium," and "high" classes for the optimum partitioning (Class SS = 537) are 10, 13, and 2, respectively.

For the Freitas et al. (1966) data of Table 1, we now have five subgroups with the following ΔY_{max} means (number of observations in parentheses):

709.1(4) ; 369.8(4) ; 242.5(3) ; 121.1(11) ; −39.4(2).

The within subgroup mean square is $s^2 = 36{,}024$; this will be used as the estimated error variance with 19 degrees of freedom. If one desires only two classes, he can compare the two class means formed by pooling successive subgroups; there will be four such contrasts. The t-statistic for any contrast is

$$t = (\bar{Y}_1 - \bar{Y}_2) \Big/ \sqrt{s^2\left(\frac{1}{n_1} + \frac{1}{n_2}\right)} \qquad [5]$$

where $\overline{Y}_i$ is the mean for the ith class and n_i the number of pooled observations in that class. Since the mean for the first class is expected to exceed that for the second class, we will use a one-tailed t-test with 19 degrees of freedom.

Before proceeding to determine the optimal number of classes, we first test if adjacent subgroups have significantly different means. The t-test for 1 versus 2 is

$$t_{12} = (709.1 - 369.8) \Big/ \sqrt{s^2(\frac{1}{4} + \frac{1}{4})} = 2.53$$

which is significant at the 1% level. Since subgroups 2 and 3 are not significantly different ($t = 0.88$), they are pooled in subsequent analyses. Similarly subgroups 4 and 5 are pooled because $t = 1.10$. This leaves three classes with means and numbers of observations as follows:

$$709.1(4) \ ; \quad 315.2(7) \ ; \quad 96.4(13) \ .$$

The division points, as in Table 1, are 32.5 ppm K and 71.5 ppm K. If one desires only two classes, the 32.5 split is optimal; however, in this case the combinations of subgroups 1 and 2 and subgroups 3, 4, and 5 with the division at 46.5 ppm K is almost as good.

It might be argued that we should use multiple comparison procedures in making these tests. However, the natural ordering of the expected values of the means would seem to validate the above sequential testing procedure.

USE OF ANALYSIS OF VARIANCE METHODS FOR THE BRAY DATA

The data from which Bray (1944) calculated the percentage yields which were used to illustrate his adaptation of the Mitscherlich equation are given in Table 2. We have refit the Mitscherlich equation in the form which we consider correct, i.e.,

$$\Delta Y_{max} = \theta_1 + \theta_2 e^{-\theta_3 X} \qquad [6]$$

where X is the soil test level and the θ_i are unknown parameters estimated by nonlinear least squares.[3] The least squares equation obtained is

$$\Delta Y_{max} = 2.136 + 62.983 e^{-.04474X} \ .$$

---

[3]The nonlinear least squares computing procedure is contained in the New Version of Statistical Analysis System (SAS), Department of Statistics, North Carolina State University and SAS Institute, Inc.

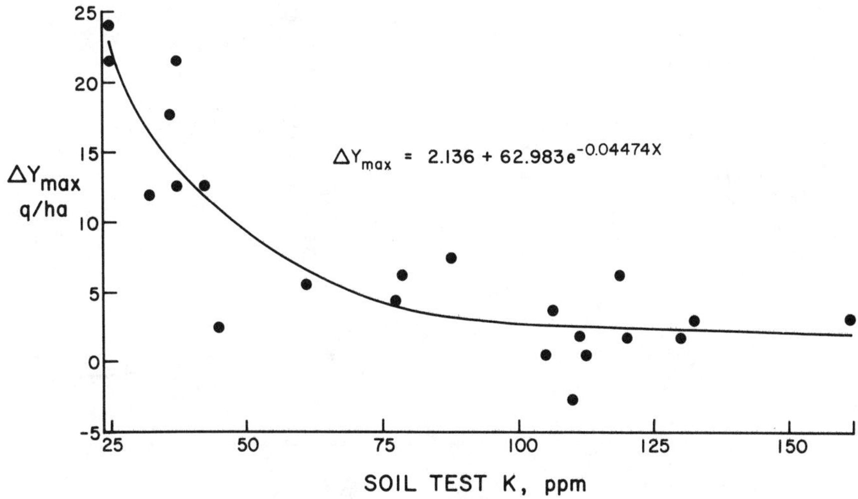

Figure 4. Scatter diagram and plot of Mitscherlich curve for Bray data.

Fitting in this manner without taking logarithms eliminates the problem caused by negative $(100 - Y)$ values discussed earlier, and the resulting biases in estimation of the coefficients. The resulting curve is shown in Figure 4.

In addition to the continuous curve, it also seemed desirable to partition the soil test axis into discrete recommendation classes such as were developed for the Freitas et al. (1966) data. The data and class sums of squares are presented in Table 2. For the modified procedure, four subgroups were created, as indicated in Table 2. The subgroup means and numbers of observations are:

$$22.66(2)\ ;\quad 15.35(5)\ ;\quad 5.10(5)\ ;\quad 2.06(10).$$

The within subgroup mean square is: $s^2 = 7.66$ with 18 degrees of freedom. One might consider having five subgroups with a division at 119 ppm K; however, this is ruled out because the mean of the last group would be larger than the previous mean. Although all adjacent means differ significantly, the contrast for the last two is barely significant ($t = 2.01$); hence, if only three classes are desired, one should pool these two subgroups to produce a mean of 3.07 with 15 observations. The division points would be, as indicated by the Analysis of Variance, at 29 ppm K and 44 ppm K. If only two classes are desired, the best division point is at 44 ppm K.

USE OF ANALYSIS OF VARIANCE METHODS FOR THE ARNOLD-SCHMIDT DATA

The two and three class models were also fitted to the tomato (*Lycopersicon esculentum*) yield data of Arnold and Schmidt (1951) which

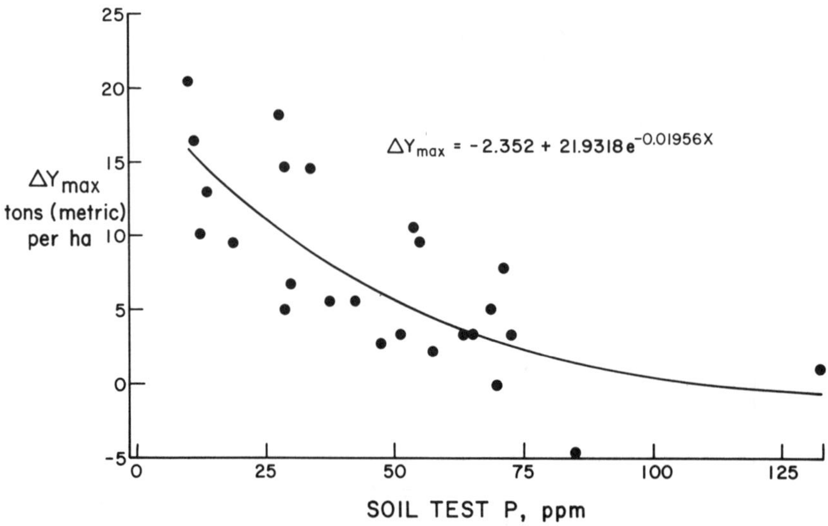

Figure 5. Scatter diagram and plot of Mitscherlich curve for Arnold-Schmidt data.

had been fitted by the authors using a Bray-Mitscherlich model. In addition, we refit the Mitscherlich model using Eq. [6]. The resulting equation is $\Delta Y_{max} = -2.352 + 21.9318e^{-.01956X}$, the graph of which is shown in Figure 5. Class Sums of Squares for various two and three class partitions are reported in Table 3. There are five subgroups with means and numbers of observations as follows:

$$18.03(2)\ ;\quad 11.38(8)\ ;\quad 6.23(6)\ ;\quad 3.65(7)\ ;\quad -1.68(2).$$

The estimated error variance is $s^2 = 12.49$ with 20 degrees of freedom.

The first three subgroups differ significantly; the last two subgroups do not. Since the third and fourth subgroups differ least, they should be pooled to produce a mean of 4.84 with 13 observations. Now the means of all four subgroups differ significantly. Since the last subgroup indicates no effect of added P, we recommend omitting it from further discussion. (Soils with more than 80 ppm P should not receive added P). Hence for three classes, use division points at 12.5 ppm P and 36 ppm P with class means of 18.03, 11.38, and 4.84, based on 2, 8, and 13 respective observations. If only two classes are used, we would combine the first two subgroups (lower value of t than for other two), giving a division point of 36 ppm P with means of 12.70 and 4.84, based on 10 and 13 respective observations.

SUMMARY OF RESULTS FOR THE THREE DATA SETS

A three-class model provided a better fit than the two-class model for the three sets of data given in Tables 1, 2, and 3. One partition was common between the two and three class models for each set of data. This led us to the conclusion that these breaks may be natural and not an artifact of the point configuration. Two of the data sets (Bray, 1944; Arnold & Schmidt, 1951) had only two observations in one of the classes. More observations should be obtained in the regions having such a limited number of points in order to verify the partitions.

The scatter configuration and graphical results had previously led us to believe that two classes would adequately fit the Freitas et al. (1966) data. After establishing that three classes should be used for these data, we are now inclined to reanalyze a number of Latin American data sets which had similar scatter diagrams and which were also assumed to be two-class situations.

Both graphical and analytical results indicate that percentage yields for these three sets of data have lower variance than ΔY_{max}. Assume we have a model

$$Y_i = \mu_i + b + e_i, \qquad i = 0, 1$$

where Y_i refers to Y_{max}, μ_i is the mean of Y_i, b is a random environmental effect common to both treatments (variance σ_b^2) and e_i is the random environmental effect not common (variances, σ_0^2 and σ_1^2). Then the approximate squared coefficient variation for $p = Y_0/Y_{max}$ is

$$(CV)_p^2 \doteq \frac{\sigma_0^2 + \sigma_b^2}{\mu_0^2} + \frac{\sigma_1^2 + \sigma_b^2}{\mu_1^2} - \frac{2\sigma_b^2}{\mu_0\mu_1}$$

$$= \frac{\sigma_0^2}{\mu_0^2} + \frac{\sigma_1^2}{\mu_1^2} + \sigma_b^2\left(\frac{1}{\mu_0} - \frac{1}{\mu_1}\right)^2 .$$

Similarly for ΔY_{max}

$$(CV)_\Delta^2 = \frac{\sigma_0^2 + \sigma_1^2}{(\mu_1 - \mu_0)^2} \quad .$$

If $\sigma_0^2 = \sigma_1^2 = \sigma^2$ (a reasonable assumption) and $\sigma_b^2 = \rho\sigma^2$

$$(CV)_p^2 = \sigma^2\left[\left(\frac{1}{\mu_0^2} + \frac{1}{\mu_1^2}\right) + \rho\left(\frac{1}{\mu_0} - \frac{1}{\mu_1}\right)^2\right]; \qquad (CV)_\Delta^2 = \frac{2\sigma^2}{(\mu_1 - \mu_0)^2} .$$

If $\mu_1 = 2\mu_0$, $(CV)_p^2 = (5 + \rho)\sigma^2/4\mu_0^2$, and $(CV)_\Delta^2 = 2\sigma^2/\mu_0^2 > (CV)_p^2$, unless $\rho \geqslant 3$. However, if $\mu_1 = 3\mu_0$, $(CV)_p^2 = (10 + 4\rho)\sigma^2/9\mu_0^2$, and $(CV)_\Delta^2 = \sigma^2/2\mu_0^2 < (CV)_p^2$. As μ_1 approaches μ_0, ΔY_{max} will become relatively much more variable than p; however, when μ_1 is much larger than μ_0, ΔY_{max} becomes relatively less variable than p. For the range of data in Tables 1, 2, and 3, the coefficient of variation of percentages will usually be less than for ΔY_{max}. It is clear that ΔY_{max} removes more site variation than does the percentage.

PARTITIONING OF MULTINUTRIENT RESPONSE DATA

The joint dependence of crop yield on more than one soil nutrient implies that soil test correlation should be carried out on two or more nutrients simultaneously. It is especially important to use such an approach where interaction between the added nutrients occurs. One difficulty is that interaction often occurs only at some sites of the series. This makes modeling of the response versus soil test level more difficult.

Most of the soil test correlation data in the literature involving enough sites to adequately evaluate the correlation are for a single nutrient. Other data sets, not specifically designed for soil test correlation, give two nutrient yield response results but the soil test information is not reported. Many data sets involve nitrogen as the principal nutrient to which response is obtained but correlations involving yield and soil test nitrogen are usually not strong due to difficulties in evaluating this nutrient by soil test.

The ideal data for developing multinutrient correlation methods would be from a series of P × K factorial experiments. To develop methodology, one should have a series of sites in which response is limited to both P and K and for which there is an interaction between these added nutrients at most locations. This would facilitate building the interaction complexity into the model.

Bray (1948) studied P and K correlation simultaneously using the percentage yield variable. However, he did not report the data or give details of the correlation procedure. It was reported, however, that the percentage yield was calculated using Baule's percentage sufficiency concept. If the percentage yield for P was 70% and that for K was 80%, the combined percentage yield for the two nutrients was $(0.80 \times 0.70) \times 100 = 56\%$. Such an approach makes rather rigid assumptions about the nature of interaction which is known to be quite unpredictable in many cases. One should allow for the possibilities of no interaction or negative interactions among the fertilizer nutrients.

Percentage yield has the same problems noted for the single nutrient case, namely, lack of direct agronomic and economic relevance. Also, if there is an interaction, there exists a problem in deciding what to use as the denominator in computing percentage yield. Main effect means could be used for calculating percentage yield in the absence of interaction.

Again, for the multinutrient case, we recommend using ΔY_{max} instead of percentage yield as the yield response variable. There would be three possible ΔY_{max}'s, i.e., ΔY_{maxP}, ΔY_{maxK}, and ΔY_{maxPK}. The latter would apply only in cases where interaction occurs. The independent variables would be soil test P, soil test K and soil test P × soil test K. The coded levels of the independent variables would be 0 and 1 for the "low" and "high" levels, respectively. Arbitrary partitions between "low" and "high" could be set up for each of the nutrients and the Cell Sum of Squares determined for each combination of partitions for the two nutrients. Coding of the levels of P and K for using multiple regression techniques to carry out the calculations would be as shown below.

		P Low	P High
K	Low	00	01
K	High	10	11

The number of observations in the four cells would vary depending upon the partitions chosen for P and K. The best combination of partitions would be the one which maximizes the Cell Sum of Squares (SS Regression).

Cluster analysis, as described by Beale (1969), could be used on the ΔY_{max} variable as an alternative to the procedures described above. It is an exploratory technique for doing in *n*-dimensions the things done with graphical techniques on scatter diagrams in two dimensions.

Another approach is to incorporate soil test results for more than one nutrient into the "multivariate yield response model" discussed in the section on Previous Models. There is considerable flexibility for including terms representing soil test variables, including interactions, into the model. The limitations of this approach are discussed in the Previous Models section.

There is a definite need for someone to conduct a series of two-nutrient experiments for the specific purpose of multinutrient correlation, perhaps using some of the techniques described here. Meanwhile, the effects of other nutrients should be recognized and given some consideration in interpreting single nutrient correlation results.

LITERATURE CITED

Anderson, R. L., and L. A. Nelson. 1975. A family of models involving intersecting straight lines and concomitant experimental designs useful in evaluating response to fertilizer nutrients. Biometrics 31:303-318.

Arnold, C. Y., and W. A. Schmidt. 1951. Soil tests as a measure of phosphorus available to tomatoes on heavy soils. Soil Sci. 71:105-115.

Beale, E. M. L. 1969. Euclidean cluster analysis. Bull. Int. Stat. Inst. 43 (Book 2) 92-93.

Bray, R. H. 1944. Soil-plant relations: I. The quantitative relation of exchangeable K to crop yields and crop response to potash additions. Soil Sci. 58:305-324.

Bray, R. H. 1948. Correlation of soil tests with crop response to added fertilizers and with fertilizer requirement. p. 53-86. *In* H. B. Kitchen (ed.) Diagnostic techniques for soils and crops. American Potash Institute, Washington, D. C.

Cate, R. B., Jr., and L. A. Nelson. 1965. A rapid method for correlation of soil test analyses with plant response data. North Carolina Agric. Exp. Stn., International Soil Testing Series Tech. Bull. No. 1.

Cate, R. B., Jr., and L. A. Nelson. 1971. A simple statistical procedure for partitioning soil test correlation data into two classes. Soil Sci. Soc. Am. Proc. 35:658-659.

Colwell, J. D. 1967. The calibration of soil tests. J. Aust. Inst. Agric. Sci. 33:321-330.

Fitts, J. W. 1955. Using soil tests to predict a probable response to fertilizer application. Better Crops Plant Food. 39:17-20.

Fitts, J. W., and W. L. Nelson. 1955. The determination of lime and fertilizer requirements of soils through chemical tests. Adv. Agron. 8:241-282.

Freitas, L. M. M. de, A. C. McClung, and F. Pimentel Gomes. 1966. Determination of potassium deficient areas of cotton. Potash Rev. 5-25:1-12.

International Soil Fertility Evaluation and Improvement Program. 1971. Annual Report, 1971. North Carolina State University, ISFEI Program, Raleigh.

Laird, R. J., and F. B. Cady. 1969. Combined analysis of yield data from fertilizer experiments. Agron. J. 61:829-834.

Olmstead, P. S., and J. W. Tukey. 1947. A corner test of association. Ann. Math. Stat. 18:495-513.

Soil Test Work Group of National Soil Research Committee. 1956. Soil tests compared with field, greenhouse, and laboratory results. North Carolina Agric. Exp. Stn. Tech. Bull. 121.

Spillman, W. J., and E. Lang. 1924. The Law of diminishing returns. World Book Company, Chicago.

Contrasting Concepts in Soil Test Interpretation: Sufficiency Levels of Available Nutrients Versus Basic Cation Saturation Ratios

E.O. McLean[1]

Concepts in soil test interpretation differ greatly. All of them are based on valid principles. Any given one seems to work well under specific conditions, but insufficient research has been done to make one concept applicable to all conditions. The two concepts most widely used are: i) sufficiency levels of available nutrients (SLAN) and ii) basic cation saturation ratios (BCSR). Both concepts recognize that there is a fraction of a given plant nutrient measured by soil tests which is readily available to plants, and that the level of available nutrient measured may range from low to high in a given soil.

The SLAN concept, is based on a general mathematical expression of the law of diminishing returns where increases in yield of a crop per unit of available nutrient decreases as the level of available nutrient approaches sufficiency. As originally conceived, this expression was thought to apply to all crops and to all soil conditions, but early work showed it had to be adjusted for different crops (Bray, 1958). Likewise, the BCSR concept which holds that for most crops there is a best ratio of basic cations on the soil cation exchange capacity (CEC), is found not to be as simple as originally thought.

Marked shifts from large usage of the SLAN concept to much greater usage of the BCSR concept as bases for recommending fertilizer and lime have occurred in recent years. In 1960, a total of 1.25 million samples were tested in the north-central region of the U.S. of which university laboratories tested two-thirds and nonuniversity laboratories tested the remainder. By the end of the 1960's a total of 2 million samples were tested in the same region, with university laboratories testing only one-fifth of the total

[1]Professor, Ohio State University and Ohio Agricultural Research and Development Center, Columbus.

(Darst, 1970). Also, essentially all of the commercial and industrial laboratories use the BCSR concept while 11 of the 12 university laboratories use the SLAN concept. Hence as commercial and industrial laboratories have taken over a larger share of the soil testing business, a correspondingly larger share of the recommendations are based on the BCSR concept.

OBJECTIVES IN SOIL TESTING

The aims in soil testing are to assess the relative adequacies of available nutrients (or lime requirements) and to provide guidance on amounts of fertilizers (or lime) required to obtain optimum growth conditions for plants. Such objectives are contingent upon securing representative samples of soil, analyzing them in an acceptable manner, and calibrating the effects on crop response (or increased soil test values) of added units of fertilizer (or lime) to soils at various initial levels of available nutrients. Such calibrations, as make interpretations of soil tests possible, are what this paper is about.

The reader may wish to refer to the excellent general discussions of correlation, calibration and interpretation of soil test results by Hanway (1973), Cope and Rouse (1973), and Barber (1973) in the book, *Soil Testing and Plant Analysis*.

MECHANISMS REGULATING REACTIVITY OR PLANT-AVAILABILITY OF NUTRIENTS IN SOILS

As previously implied, only a relatively small fraction of most plant nutrients contained in a soil are in solution or in an otherwise reactive form. The ions in solution are in equilibrium with those held in solid or nonactive form by several mechanisms (McLean, 1972). Each mechanism which inactivates ions but gradually releases a fraction of them into the soil solution as their concentration is lowered by various means can be regarded as an ionic reservoir (Fig. 1). These ionic reservoirs are: cation exchange capacity (CEC), anion exchange capacity, chelation or complexation capacity, solubility-insolubility capacity, microorganismic absorption, ion pair formation, and possibly others. All of these mechanisms (ionic reservoirs) bind the ions at some level of bonding energy. The stronger an ion is bonded, the lesser its tendency to go into solution and the lower its availability to plants. The mechanism which restricts the availability of a given ion most depends upon: (i) the relative magnitude of the mechanism itself for regulating a given ion, (ii) the relative amounts of the ion, and (iii) the relative strengths of bonding between mechanism and ion. For example, due to the relatively large magnitudes of the CEC of most soils compared to the other mechanisms, and of the existing amounts of exchangeable Ca, the CEC mechanism is the one most often involved in regulating the availability of Ca to plants. On the other hand, a cation such as Cu may be regulated more by the chelation mechanism, because of its smaller quantity and stronger bonding by chelation than by CEC. Phos-

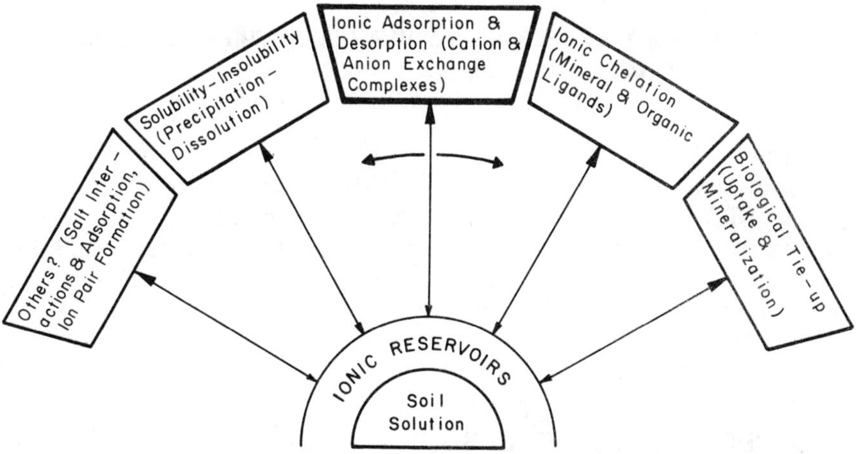

Figure 1. Schematic diagram of the ionic reservoirs (mechanisms) removing ions from and returning ions to the soil solution.

phorus is regulated in part by adsorption of the phosphate ion to the anion exchange mechanism, and in part by the relative insolubility of the phosphate ion in the presence of cations such as Fe, Al, and even Ca. A test designed to measure the amount of an element which is plant-available in a soil will, of course, remove that in solution as well as a sizable fraction of the more reactive form regardless of which mechanism(s) bond(s) it.

It should now be evident that of the several mechanisms (ionic reservoirs) mentioned above, the BCSR concept embraces only the CEC and its adsorption of the basic cations. Also, this concept deals with only one of the major fertilizer elements, i.e., K, and the two lime elements, Ca and Mg. It does not embrace N or P.

IONIC ENVIRONMENT: OPTIMUM CONDITIONS FOR PLANT GROWTH VERSUS ADEQUATE LEVELS OF NUTRIENTS (OR LOWEST LEVELS OF TOXIC SUBSTANCES)

The basic cation saturation ratio (BCSR) concept in soil test interpretation is based on an "idealized" procedure for providing an ionic environment with optimum conditions for plant growth. In practice, the optimal condition may embrace either such a broad range of cation ratios that it can be identified only by sophisticated procedures, or it may differ with each major soil condition. Certainly, a plant often tolerates a considerable range: (i) in level of available nutrients or (ii) in level of toxic substances without appreciable effect on yield or quality of harvested product. Also, with appropriate selection of the specific cultivar, plants can be found which will tolerate almost any soil condition. However, the BCSR concept seeks to attain optimal conditions for a wide range of crops, while the SLAN concept focuses on the adjustment of the one or more elements known to be at less-than-sufficiency levels for a given crop.

SUFFICIENCY LEVELS OF AVAILABLE NUTRIENTS (SLAN) CONCEPT IN SOIL TEST INTERPRETATION

Implications of the SLAN Concept

This concept implies: (i) that levels of available nutrients range in a group of soils from insufficiency to sufficiency for optimum growth of plants, (ii) that amounts of nutrients removed by suitable extractants will be inversely proportional to yield increases from added nutrients, and (iii) that calibrations have been made for changing the levels of available nutrients in the soil by adding fertilizer (or lime).

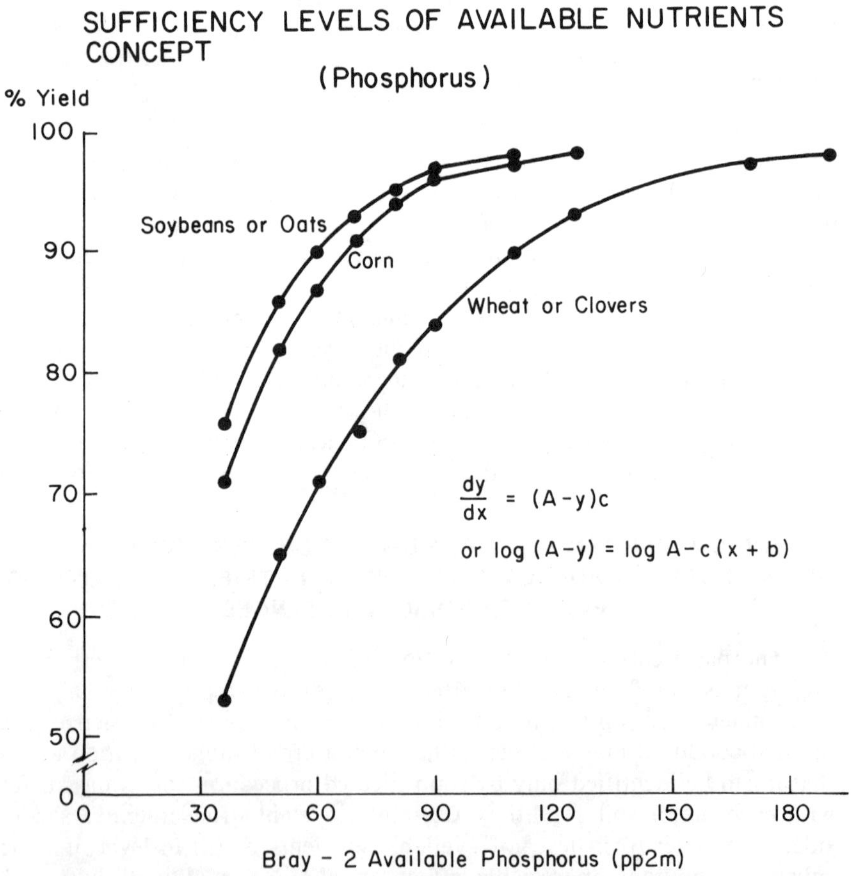

Figure 2. Percentage yield of various crops according to level of Bray-2 available phosphorus in parts of phosphorus per 2 million parts of soil. Sufficiency level = 100% yield.

Origin of the SLAN Concept

This concept resulted from research on crop response to addition of an element otherwise insufficient for maximum crop yields (Bray, 1944, 1945). The response was observed to be progressively smaller with each succeeding increment of the growth-limiting element until further response was zero or negative with additional increments of element (Fig. 2 and 3). Analysis of these incrementally-treated soils provided calibrations of change in level of available nutrient per unit of added element. Also, the levels of available nutrient in the incrementally-treated soils were correlated with the yields of the crop.

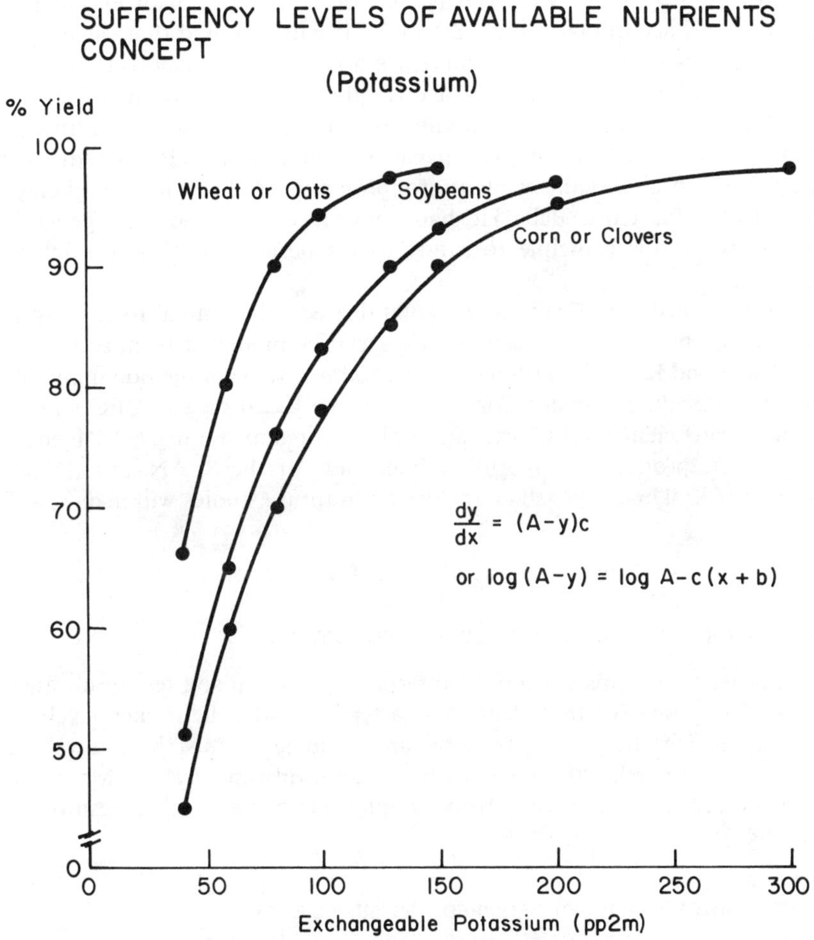

Figure 3. Percentage yield of various crops according to level of exchangeable potassium in parts of potassium per 2 million parts of soil. Sufficiency level = 100% yield.

Principles Involved in the SLAN Concept

This concept assumes little, if any, effect of the level of availability of one ion on that of another. However, it does recognize that an addition of the most limiting element may cause more efficient utilization of a less-limiting element. The concept is based on Mitscherlich's equation: $dy/dx = (A - y)c$ which says in effect that yield increases (dy) per unit of available nutrient (dx) decrease as the current yield (y) approaches a maximum yield (A), c being a proportionality constant (Bray, 1944). Contrary to Mitscherlich's hypothesis that c was a constant for all crops and soil conditions, Bray (1945) found that the proportionality constant for available P and exchangeable K was different for different crops or groups of crops. Hence the curves for percentage yield versus level of available P or of exchangeable K differed accordingly (Figs. 2 and 3). It will be noted that the crops which group together for a given curve are not the same for both P and K. Since Bray found that the values for c for these crops or groups of crops were relatively constant in the medium- and fine-textured soils of Illinois, he felt that a given level of exchangeable K or available P was equally available to a specific plant or plant group on almost all silt loam and clay loam soils of the Corn Belt. He had some reservations on the general applicability of this principle to sandy loams because of their droughty nature.

Recently, Fisher (1975) used a second degree polynominal to express a fertility index by which a percentage yield can be predicted from soil tests values for P and K. He substituted for the soil test value in the polynominal equation for K an expression for soil test value based on soil CEC which provides a sufficient level of exchangeable K for maximum yields. Hence he, in effect, produced a hybrid or intergrade of the SLAN and BCSR concepts for K. This may well be the direction future studies will take.

Steps in Usage of the SLAN Concept

SELECTION OR DETERMINATION OF SUFFICIENCY LEVEL

In order to use this concept of interpreting a given soil test result and recommending rates of that element as a fertilizer, the sufficiency level of available nutrient for a specific crop and yield level must be known or estimated from results of studies with this crop on similar soils. Many of these values in use today come from adaptations of the original results of Bray (1945).

COMPUTATION OF FERTILIZER REQUIRED FOR SUFFICIENCY

Although Bray (1958) noted that soluble phosphates rapidly convert to adsorbed form and that the proportionality factor for the adsorbed form of P in the Mitscherlich equation appeared to be constant for all non-calcareous soils of the Midwest, most workers now agree that the degree of fixation and the types of P formed vary considerably. Hence the amount of soluble P to raise the available P from the initial to the sufficiency level will differ with the initial level and with soil conditions such as texture and degree of weathering. Most accurate values may be obtained only when research covering a range in conditions has been completed.

METHODS OF SUPPLYING THE FERTILIZERS (AND LIME)

The element(s) to be added may be applied in various ways. Two widely used ones are: (i) soil build-up plus crop needs and (ii) crop needs plus partial build-up. The first of these emphasizes correcting deficiencies in the soil and making certain that the crop does not lower the availability of the elements below the sufficiency levels established. The second method emphasizes applying sufficient elements to meet crop needs plus small amounts extra to build-up the soil.

Soil Build-up Plus Crop Needs—As mentioned, this method of application of fertilizer corrects deficiencies in availability of the various elements immediately, and subsequently adds sufficient amounts of the elements annually to maintain them at the sufficiency levels. This method of applying fertilizer is probably most practical for the farm owner who expects to remain in business over a period of years and thus can afford the initial investment from which the highest possible yields are likely to be obtained. Economies in purchase and application may be effected as a consequence of the higher initial rates being applied.

Crop Needs Plus Partial Build-up—As mentioned, this method of application of fertilizer provides for crop needs plus an amount left over to build up the soil. The latter amount may be some fixed fraction such as one-fifth or one-tenth that required to reach sufficiency levels for the elements. Although the yields of crops from this method of supplying nutrients may not quite reach maximum, they will likely be nearly so. This method of application is probably most practical for the farm operator who must live with short-term leases which do not provide time for recovery of fertilizer when applied at rates required to reach the full level. However, the method may be applicable to almost everyone during periods of fertilizer shortage or extremely high cost.

Bray (1958) noted that when climatic conditions favored higher yields, treated and check plot yields both increased correspondingly. Hence he found no evidence that the yield level should be taken into consideration in determining degree of sufficiency (or level of fertilizer to apply) for immobile nutrients such as P and K. He recognized, however, that it was widely done for N, P, and K, but he felt this should be done only for the mobile nutrient, N.

BASIC CATION SATURATION RATIO (BCSR) CONCEPT IN SOIL TEST INTERPRETATION

Implications of the BCSR Concept

This concept implies that for optimum growth of crops both a best ratio of basic cations and a best total base saturation exist in a soil. However, little data have been published to confirm either that a best basic cation saturation ratio (and total base saturation) exists for all soils, or that it should vary from one soil to another. Although not strictly a part of the BCSR concept, its strong emphasis on an idealized plant environment has led most laboratories using the concept to include tests for secondary elements. Where these test results indicate a likelihood of crop response to one or more of these elements, they are included in the fertilizer (and lime) recommendations. This is as it should be. However, overzealous people often make recommendations of secondary elements for soils not deficient in these elements, and in doing so bring criticism of laboratories using the BCSR concept. This could easily be avoided by confining such recommendations to soils shown to be deficient in one or more of these elements.

Origin of the BCSR Concept

This concept resulted from research on soils where cation saturations were varied widely and both yields and compositions of crops observed. The initial work by Bear and co-workers (1945) was done with alfalfa (*Medicago sativa* L.) on one New Jersey soil using uniform amounts of exchangeable Mg and H and variable amounts of exchangeable K and Ca. From this initial study, the following saturations of cations were selected for the "ideal" soil: 65% Ca, 10% Mg, 5% K, and 20% H (Fig. 4). Other observations by these workers (Bear et al., 1945) suggested: (i) that the 10% Mg saturation might be minimal for alfalfa rather than optimal, (ii) that soluble Mg sources may be essential for correcting Mg deficiencies in sandy soils and in those kept at low pH for growing crops like potatoes (*Solanum tuberosum*), and (iii) that liming above 80% base saturation, i.e., 20% H, (pH 6.5) brought about deficiencies of manganese and other microelements. The choice of these specific saturations was influenced as much by practical (economic) and other agronomic considerations as by crop yields. For example, they reported good growth of alfalfa through a wide range in Ca/K ratios from 100:1 to 1:1 (Hunter et al., 1943). Also, they recognized that K is taken up in much greater amounts per unit of exchangeable K than is Ca. Hence, since the cost of Ca obtained from lime was less than that of K from KCl, adequate Ca should be provided through liming in order to prevent uptake of the more costly K in greater quantities than needed by the plant. The initial studies showed that when Ca/K ratio in the tissue exceeded 4:1 ($>$ 2% Ca or $<$ 1% K content of the tissue), the yields decreased.

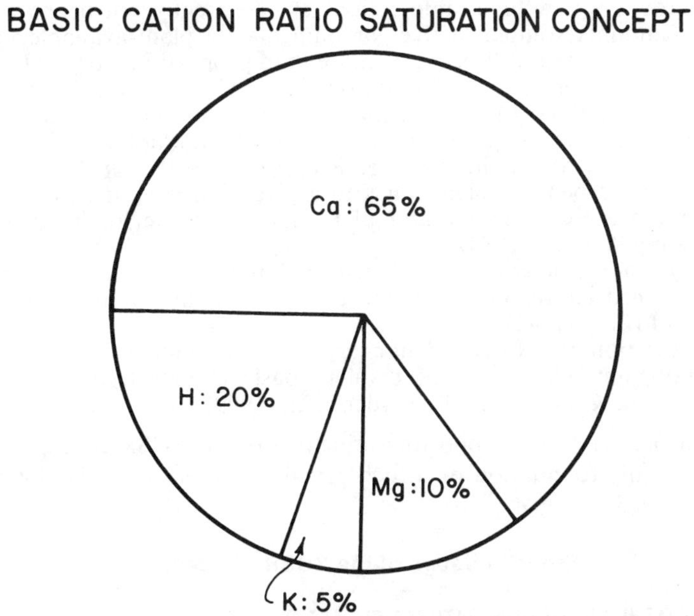

Figure 4. Schematic diagram of the percentage saturations of the cation exchange capacity of the "ideal" soil with various cations, adaptations of which lead to the BCSR concept.

The above-mentioned cation saturations of the "ideal" soil used in the initial study and the critical Ca/K ratio in the plant tissues were then used by Bear and Toth (1948) as criteria for examining 20 of the most important agricultural soils of New Jersey. The soils were adjusted to pH 6.5 with dolomitic limestone and K added at 224 kg/ha (200 lb/acre). From the results of these studies, they concluded that conditions approach the optimum for the cation nutrition of alfalfa when the soils are saturated to the levels of the "ideal" soil they had defined earlier (Bear et al., 1945). Later, Graham (1959), on the basis of unpublished work, suggested that the percentage saturations for Ca, Mg, and K be 75, 10, and 2.5 to 5%, and that ranges from 65 to 85% Ca, 6 to 12% Mg, and 2 to 5% K would not likely affect yields appreciably.

Principles Involved in the BCSR Concept

This concept arises from and is largely supported by the following principles:

1. Bonding of cations to exchange sites differs greatly from one type of cation to another (McLean & Marshall, 1948), and it differs greatly for the same type of cation at different saturations (Albrecht, 1940; Allaway, 1945; Chu & Turk, 1949; Mehlich & Colwell, 1943).

2. Amounts of exchangeable cations are frequently not proportional to amounts soluble, active, or immediately plant-available. They may be less than, directly, or more than proportional depending on the saturation level of the ion on the CEC of the soil (McLean & Marshall, 1948; McLean & Snyder, 1969), and excess of one type of cation may depress the activity and plant uptake of another (McLean & Marshall, 1948; Jenny & Ayers, 1939).
3. Ions adsorbed complementary to a given ion have marked effects on the exchangeability and plant uptake of the ion in question (Jenny & Ayers, 1939).
4. Both the capacity (total exchangeable) and intensity (activity) of an adsorbed cation influence the total availability of a cation to a plant (Bray, 1942).
5. Saturation of pH-dependent charges markedly increases the activity and plant availability of divalent basic cations (Allaway, 1945; McLean & Snyder, 1969; Snyder et al., 1969).

The aforementioned supporting principles of the BCSR concept are to some degree the foundation on which practically every topic in this publication is based.

Steps in Usage of the BCSR Concept

SELECTION OF BASIC CATION SATURATION RATIOS

As mentioned, it has been recognized (Graham, 1959) that some range in the ideal saturations may occur without adversely affecting the crop. Graham (1959) suggested the ranges: 65 to 85% Ca, 6 to 12% Mg, and 2 to 5% K. Others (Hunter et al., 1943; McLean & Carbonell, 1972) have noted that these saturations and the resultant ratios may vary much greater than the above ranges without appreciably affecting crop yields. However, one must select a specific (or a range in) basic cation saturation, and a pH level to which the soil is to be limed, if the BCSR concept is to be used as a guide in corrective fertilizer and lime applications.

DETERMINATION OF SOIL CEC

The soil is analyzed for exchangeable bases, and for lime required to bring the soil to the chosen pH. The soil CEC is obtained by totaling the basic cations and the acidity (exchangeable H and Al ions), each expressed as meq/100g of soil. Since the extracting solution dissolves some of the basic cations not adsorbed to cation exchange sites in calcareous soils, the summation of the cations extracted is not a satisfactory method for determining CEC of such soils.

COMPUTATION OF AMOUNTS OF Ca, Mg, AND K REQUIRED

Each basic cation is expressed as a percentage of the total CEC. Cations must be added at least to the extent that the existing percentage saturations of the basic cations are below the lower percentage saturations

Table 1. Additional amounts of basic cations required to increase the saturations of each as an exchangeable basic cation to the midpoint of the selected saturation range (soil CEC = 15 meq/100g)

Basic cation	Selected range	Exch. basic cations			Additional basic cations req'd		
	%	kg/ha	(lb/acre)	% Sat.	% Sat.	kg/ha	(lb/acre)
Ca	65–85	2688	(2400)	40	35	2354	(2100)*
Mg	6–12	403	(360)	10	0	0	(0)
K	2–5	213	(190)	1.6	1.9	250	(223)

* Equivalent to 5.8 metric tons/ha (2.6 tons/acre) of $CaCO_3$.

of the ranges chosen. For example, suppose the following saturation ranges are selected, and the indicated levels are present in a soil of pH 5.6 with a CEC of 15 meq/100g; then the amounts of each basic cation to be added will be as indicated in Table 1.

If more lime than the 5.8 metric tons/ha (2.6 tons/acre) were required to bring the soil to the pH selected, such as 6.5, the equivalent of an additional 10% Ca saturation 1.6 metric tons/ha (0.75 tons/acre) of $CaCO_3$, could be applied without going beyond the upper point (85%) of the selected saturation range (15 meq CEC $\times$ 0.10 = 1.5 meq Ca = 600 lb/acre = 1500 lb $CaCO_3$/acre.

Since the CEC of this soil is somewhat above average, the midpoint of the selected K range may be higher than needed as a basis for computing K to be added. A commonly used equation for level of exchangeable K desired based on soil CEC (in kg/ha) is: K = 246 + 5.6 $\times$ CEC (Fisher, 1975). This gives 117 kg/ha to be added. (K = 246 + 5.6 $\times$ 15 = 330; 330 − 213 = 117.)

APPLICATION OF THE CORRECTIVE TREATMENTS

If the soil is of low enough pH, as was true of the example above, most of the corrective treatment may be made at once with appropriate applications of fertilizers and lime (including dolomitic limestone where Mg is tending toward the lower saturation range). However, if the soil has already been limed or does not need lime, the corrective treatment may be much more restrictive. If Ca saturation is at the high end of the chosen saturation range and Mg saturation is at the low end in a soil of near-neutral pH, it may be necessary to apply Mg as a soluble form of Mg salt such as epsom salts ($MgSO_4 \cdot 7H_2O$), which is likely to be much more expensive than MgO or dolomitic lime.

SITUATIONS WHERE THE SLAN CONCEPT SEEMS TO WORK BEST

This concept seems to work best as a basis for soil test interpretation in soils of generally high, well balanced fertility, but most deficient in one element such as P, K, or Mg. It was on such soils that Bray (1945) developed his crop response curves for P and K for different crops. Under conditions not too different from those of Bray's work, some recent studies in Nebraska suggest that the SLAN concept may be superior to the BCSR concept as a practical basis for applying fertilizers (Agronomy Department, 1975. A comparison of suggested fertilizer programs obtained from several soil test laboratory services. University of Nebraska-Lincoln, Agronomy Dep. Report No. 13). These studies involved:

1) Selecting soils from five locations in Nebraska, sampling them, and having them tested:
   a) by the University of Nebraska soil testing laboratory using essentially the SLAN concept.
   b) by four commercial soil testing laboratories using the BCSR concept.
2) Applying to each of the five soils the fertilizers recommended by the various laboratories according to soil test results, and cropping with corn (*Zea mays* L.) or sugar beets (*Beta vulgaris* L.).
3) Comparing the yields of the crops from each fertilizer treatment at each location.

The laboratories using the BCSR concept included S and trace elements in their recommendations. The average cost of the fertilizers applied according to this concept for the five locations was $141.71/ha ($57.35/acre) while the average costs of the fertilizers applied according to recommendations of the laboratory using the SLAN concept for the five locations was $63.53/ha ($25.71/acre). Yet the yields of corn (three locations) and sugar beets (two locations) were the same for all fertilizer treatments at each location. This indicates that fertility applied to effect a more idealized environment may not cause a sufficient crop response to pay the cost of the adjustment. It is also quite evident that had the S and trace elements been sufficiently deficient for yield increases due to additions of these elements, the results would have been different.

SITUATIONS WHERE THE BCSR CONCEPT SEEMS TO WORK BEST

This concept seems to work best as a basis for soil test interpretation in highly weathered soils of low pH requiring relatively major adjustments in fertility. It was on such soils of medium to low CEC and medium to coarse texture that Bear and his co-workers (1945) originally developed this concept as a basis for fertilizing and liming soils to provide optimum growth conditions for alfalfa. In such soils as these, where relatively large rates of the various cationic nutrients are required for optimizing growth conditions, it seems only reasonable to apply them in accordance with the saturations and ratios indicated by the BCSR concept. Also, the BCSR concept may be quite useful in situations where it is important to maintain a fairly high level of Mg in the soil such as is done in an effort to alleviate grass tetany in livestock (Grunes et al., 1970; McLean & Carbonell, 1972).

For example, where grasses alone are grown for pasture or hay to be fed as a major source of ruminant animal diet, using a value such as 15% saturation of the CEC of the soil with Mg and holding the K saturation to 2% or less may be a reasonable way to accomplish this objective (McLean & Carbonell, 1972).

SITUATIONS WHERE A COMBINATION OF SLAN AND BCSR CONCEPTS SEEM TO WORK BEST

A combination of these two concepts seems to work best as a basis for soil test interpretation in perhaps a majority of conditions. For example, in situations where the SLAN concept is used, it is frequently noted that the level of exchangeable K (or even of P) considered sufficient for maximum growth of plants varies with the CEC of the soil (Cope & Rouse, 1973) (Table 2). Similarly, the sufficiency level of Mg often depends on the level of exchangeable K (or NH_4) (Bear & Troth, 1948; McLean & Carbonell, 1972). In other cases where one might like to use the BCSR concept for recommending fertilizer and lime, he finds the pH is already high and the CEC nearly saturated. Hence any appreciable adjustment in ratios of the cations is extremely costly and appreciable benefits to the crop may be questionable.

The adjustment of the soil test value for K based on soil CEC in the soil K versus percentage yield relationship by Fisher (1975), previously cited, is an example of a combination of the SLAN and BCSR concepts for recommending K applications for meeting crop needs.

GREATEST NEED: RESEARCH ON CROP RESPONSE TO FERTILIZERS AND LIME ADDED ACCORDING TO THE SLAN OR BCSR CONCEPT

Although each of these concepts was conceived from a reasonable amount of initial data, when one considers the number of soils being tested and the amount of fertilizer and lime being applied according to one or the other of these concepts, he is dismayed at the paucity of data for supporting developments and refinement of these concepts. Because of greater interest among soil scientists in the SLAN concept, more research has been done on crop response to fertilizers applied and interpreted according to this concept. Soils in Alabama (Cope & Rouse, 1973) are placed in one of two groups for rating soil-test values for available P. Similarly, they are placed in one of three groups for rating soil-test values for exchangeable K (see Table 2). Data in this table supplied by J. T. Cope, Jr. illustrate: (i) the ranges in soil CEC grouped together by these workers, (ii) the ranges in levels of exchangeable K considered low, medium, or high in available K at the various soil CEC values, and (iii) the change in percentage saturation of the CEC with K as the soil CEC changes. It is evident that at a given

Table 2. Exchangeable K levels, recommended rates of K for cotton (*Gossypium* sp.), and resulting total exchangeable K and percentage saturation of the cation exchange capacities (CEC) of Alabama soils according to relative amounts of available K (percentage sufficiency) and range in soil CEC (units in kg/ha)

Soil group	Range in CEC	Low K (60% sufficiency)				Med K (80% sufficiency)				High K (100% sufficiency)			
		Exch K	Recm'd K	Total	% Sat.	Exch K	Recm'd K	Total	% Sat.	Exch K	Recm'd K	Total	% Sat.
	meq/100 g												
1	0–5 (4)*	45	84	129	3.7	90	56	146	4.2	135	0	135	3.8
2	5–10 (8)*	67	84	151	2.2	134	56	186	2.7	201	0	201	2.9
3	>10 (12)*	90	84	172	1.7	180	56	236	2.2	270	0	270	2.6

* Figures in parenthesis are CEC's used for computing percentage saturations shown.

relative level of available K (or percentage sufficiency of K), the exchangeable K increases with soil CEC, but that even with the inclusion of the K recommended, the percentage saturation of the CEC decreases. This means that soils of low CEC require higher percentage saturation to produce maximum yields than do soils of high CEC. It also means that the amounts recommended differ with percentage sufficiency of K but not with level of exchangeable K within a level of percentage sufficiency.

Since soils workers in most other states have not seen fit to arrange their soils into groups according to CEC for purposes of evaluating soil tests and recommending K fertilization, this suggests that in most cases additional research is needed to determine whether or not such refinements in the SLAN concept are needed elsewhere. From the original observations of Bray (1945), it would seem logical that soil characteristics which may affect water and air relationships in the root zone might well justify unique grouping for soil test interpretation. Still other attributes of soils such as degree of fixation of either P or K may also be important bases for grouping soils with the aim of better soil-test interpretation. Yet few studies of either type have been undertaken.

As previously mentioned, a relatively high proportion of north-central regional soils are tested, and fertilizer and lime recommendations made by laboratories using the BCSR concept. Since many of the larger laboratories doing business in the north-central region are located in the South and West, it seems likely that more than half of all soils tested in the U.S. use this concept for interpretation of soil test results. Yet research which would indicate how basic cation saturations should vary with specific soil conditions, or how widely saturations may vary without adverse effects on yields and composition of crops, simply has not been done.

The need is great for initiating such studies, and everyone seems to agree that they must be done in the field to be meaningful in soil test interpretation (Bray, 1945; Cope & Rouse, 1973; Hanway, 1973; Thomas & Peaslee, 1973). In this connection, we have recently initiated two types of field studies which we hope will throw some light on this subject for Ohio conditions. They are: (i) levels of K for fine-textured (high CEC) soils using both broadcast and banded treatments, and (ii) widely ranging Ca/Mg/K ratios for soils at different base saturations. Many more studies of these kinds are needed for different soil and climatic conditions. Without the results of such studies, soil test interpretations as translated into fertilizer and lime recommendations will likely remain essentially correct for general conditions but may be wrong for many specific conditions.

LITERATURE CITED

Albrecht, W. A. 1940. Adsorbed ions on the colloidal complex and plant nutrition. Soil Sci. Soc. Am. Proc. 5:8-16.

Allaway, W. H. 1945. Availability of replaceable Ca from different types of colloids as affected by degree of Ca saturation. Soil Sci. 59:207-217.

Barber, S. A. 1973. The changing philosophy of soil test interpretations. p. 201-211. *In* L. M. Walsh and J. D. Beaton(ed.) Soil testing and plant analysis. Soil Sci. Soc. of Am., Madison, Wis.

Bear, F. E., A. L. Prince, and J. L. Malcolm. 1945. Potassium needs of New Jersey soils. N.J. Agric. Exp. Stn. Bull. 721.

Bear, F. E., and S. J. Toth. 1948. Influence of Ca on availability of other soil cations. Soil Sci. 65:69-74.

Bray, R. H. 1942. Ionic competition in base-exchange reactions. J. Am. Chem. Soc. 64:954-963.

Bray, R. H. 1944. Soil-plant relationships: I. The quantitative relation of exchangeable K to crop yields and to crop response to potash additions. Soil Sci. 58:305-324.

Bray, R. H. 1945. Soil-plant relationships: II. Balanced fertilizer use through soil tests for K and P. Soil Sci. 60:463-473.

Bray, R. H. 1958. The correlation of a P soil test with the response of wheat through a modified Mitscherlich equation. Soil Sci. Soc. Am. Proc. 22:314-317.

Cope, J. T., Jr., and R. D. Rouse. 1973. Interpretation of soil test results. p. 35-54. *In* L. M. Walsh and J. D. Beaton (ed.) Soil testing and plant analysis. Soil Sci. Soc. of Am., Madison, Wis.

Chu, T. S., and L. M. Turk. 1949. Growth and nutrition of plants as affected by degree of base saturation of different types of clay minerals. Mich. Agric. Exp. Stn. Tech. Bull. 214.

Darst, B. C. 1970. ASCS certification for commercial and private soil testing laboratories. *In* R. A. Weise (compiler) Proceedings Third Soil-Plant Analyst Workshop. Univ. of Nebraska, Lincoln.

Fisher, T. A. 1975. Some considerations for interpretation of soil tests for P and K. Mo. Agric. Exp. Stn. Res. Bull. 1007.

Graham, E. R. 1959. An explanation of theory and methods of soil testing. Mo. Agric. Exp. Stn. Bull. 734.

Grunes, D. L., P. R. Stout, and J. R. Brownell. 1970. Grass tetany of ruminants. Adv. Agron. 22:331-374.

Hanway, J. J. 1973. Experimental methods for correlating and calibrating soil tests. P. 55-66. *In* L. M. Walsh and J. D. Beaton (ed.) Soil testing and plant analysis. Soil Sci. Soc. of Am., Madison, Wis.

Hunter, A. S., S. J. Toth, and F. E. Bear. 1943. Calcium-potassium ratios for alfalfa. Soil Sci. 55:61-72.

Jenny, H., and A. D. Ayers. 1939. The influence of the degree of saturation of soil colloids on the nutrient uptake by roots. Soil Sci. 48:443-459.

McLean, E. O. 1972. Base saturation ratio versus insufficiency-sufficiency values as bases for fertilizing and liming of soils. Commun. Soil Sci. Plant Anal. 3:141-149.

McLean, E. O., and M. D. Carbonell. 1972. Calcium, Mg, K saturation ratios in two soils and their effects upon yields and nutrient contents of German millet and alfalfa. Soil Sci. Soc. Am. Proc. 36:927-970.

McLean, E. O., and C. E. Marshall, 1948. Reciprocal effects of Ca and K as shown by their cationic activites in montmorillonite. Soil Sci. Soc. Am. Proc. 13:179-182.

McLean, E. O., and G. H. Snyder. 1969. Interactions of pH-dependent and permanent charges of clays: I. Use of specific ion electrodes for measuring Ca and Rb activities in bentonite and illite suspensions. Soil Sci. Soc. Am. Proc. 33:388-391.

Mehlich, A., and W. E. Colwell. 1943. Influence of nature of soil colloids and degree of base saturation on growth and nutrient uptake by cotton and soybeans. Soil Sci. Soc. Am. Proc. 8:179-184.

Snyder, G. H., E. O. McLean, and R. E. Franklin. 1969. Interactions of pH-dependent and permanent charges of clays: II. Calcium and Rb bonding to bentonite and illite suspensions-clay-phase retention. Soil Sci. Soc. Am. Proc. 33:392-396.

Thomas, G. T., and D. E. Peaslee. 1973. Testing soils for phosphorus. p. 115-132. *In* L. M. Walsh and J. D. Beaton (ed.) Soil testing and plant analysis. Soil Sci. Soc. of Am., Madison, Wis.

Ion Activities and Ratios in Relation to Corrective Treatments of Soils[1]

D. E. Baker[2]

The soil test analysis aspect of soil and fertilizer management is of paramount importance to efficient crop production. Other major parameters of crop production include selection of variety or hybrid; phenology; soil conservation and water management; fertilizer, manure, and lime placement and time of application; and many cultural and management practices which affect the production capability of the soil type under consideration. Soil properties associated with soil type and production capability include depth, drainage, structure, and colloidal composition as they relate to air-water relations, and plant nutrient retention, release, and movement.

Soil testing service programs generally interpret the analytical data with the assistance of information supplied by the grower or predicted from climatic records, soil maps, and yield response data to determine yield potential. Since it is the fertilizer recommendation that the grower uses in management, soil testing is generally looked upon as a "test" to determine fertilizer and lime needs for the crop under consideration.

Commercial fertilizers, lime, manure, sewage sludge, and other sources of essential and nonessential elements for plants, animals, and human beings should be applied for the benefit of society as well as the farmer. The benefits include efficient crop production, improved environmental quality, quality of feed and food, and energy production. Changes in soil test levels are important only as they affect the kinds and rates of materials to be added in the following years. Surely we have advanced soil testing and soil fertility technology to the point where it is not necessary to debate the merits of fertilization to meet soil requirements versus crop needs. We do both as required for the benefit of all. Depending upon the yield potential of a soil in relation to the initial soil test level, the amount

---

[1]Contribution from the Pennsylvania Agric. Exp. Stn., University Park. Paper No. 4913 in the Journal Series.

[2]Professor of soil chemistry, Department of Agronomy. Acknowledgments are made to Richard Eshelman and C. E. Marshall for their suggestions and critical reviews of the manuscript.

of a particular element to be applied will range from none to much more than will be removed by the crop.

The amount of an element to be added to a soil for the growth of a crop is a function of crop removal in relation to the following:

(a) Plant availability of a specific ion as affected by adsorption and dissolution of soil minerals and/or man-induced changes from applications of fertilizers and other materials (manure, sewage sludge).
(b) Effects that other ions have on the uptake of the ion under consideration.
(c) Changes in (a) and (b) to be expected over the growing season.
(d) Differential soil-plant interactions such as soil volume per plant (Baker & Woodruff, 1963) and plant species or variety.

The objective of this presentation is to discuss the concepts of ion activities and ratios as they relate to factors affecting the amounts of an element to be applied to soils. The hypothesis to be defended is that ionic activities and ratios have advantages over other methods for measuring items a and b. With appropriate assumptions, a soil test has been designed to include all three parameters.

THEORETICAL CONSIDERATIONS

For a dissolved solute within a system, its partial molal Gibbs free energy, $\overline{G}$, is equal in all phases at equilibrium. Thus, if we have a measure of $\overline{G}$ for any phase (solution, adsorbed, or labile quantity of the solute), then $\overline{G}$ at equilibrium is defined even if the concentration varies greatly among phases. From the concept of relative fugacity (escaping tendency) in relation to the relative partial molal free energy, the activity, $\overline{a}$, is defined when the standard state activity is taken as unity, $\overline{a}^\circ = 1$:

$$\overline{G} - \overline{G}^\circ = RTln \frac{\overline{a}}{\overline{a}^\circ} \qquad [1]$$

where $\overline{G} - \overline{G}^\circ$ is defined as the relative partial molal free energy (partial molal free energy in the system, $\overline{G}$, relative to that of the standard state, $\overline{G}^\circ$), R is the universal gas constant, T is the temperature in degrees Kelvin, and ln is the natural logarithm. The partial molal free energy of an electrolyte, HCl for example, is equal to the sum for the component ions:

$$\overline{G}_{HCl} = \overline{G}_{H} + \overline{G}_{Cl}. \qquad [2]$$

For a two component solution, in which the ions are evenly distributed as in dilute solution of HCl, it is possible to measure $\overline{G}_{HCl}$ from the pH measured with the hydrogen electrode or the conventional glass electrode pH meter where at 25 C:

$$(\overline{G} - \overline{G}^\circ)_{HCl} = 2RTln\ \overline{a}_{H^+} = -\ 2728 \text{ pH calories/mole} \qquad [3]$$

For substances dissolved in a soil or colloidal suspension, the component ions are not evenly distributed, but if it is assumed that $\bar{a}_{Cl^-}$ (the suspension activity of Cl^-) is the same in all systems being compared and if it is further assumed that the distribution of Cl^- is not substantially different among soils tested, then

$$\Delta\bar{G}_{HCl} \cong \Delta G_{H^+} . \quad [4]$$

But as pointed out by Baker and Low (1970), $\Delta\bar{G}_{HCl} = \Delta(\bar{G} - \bar{G}^\circ)_{HCl}$ because $\bar{G}^\circ_{HCl}$, the partial molal free energy of HCl in its standard state, is also constant. As pointed out in 1929 by Guggenheim (Pitzer & Brewer, 1961, p. 351), the escaping tendency or activity of H^+ or any charged species is affected by the net electrical charge of the phase in which it is located. However, whatever the effect that such a net charge may have upon the escaping tendency of the H^+ ion, it will have exactly the opposite effect on the escaping tendency of the Cl^- ion. Thus, the sum of the partial molal free energies remains constant as defined in Eq. [2].

In order to emphasize the additional effect of a net electrical charge upon the escaping tendency of ions, Guggenheim used the terms "electrochemical potential" and "absolute activity," while Low (1950, 1954) defined "total potential" and "total activity" expressed in terms of the partial molal free energy which is the same everywhere within a system at equilibrium. Presentations by Bolt (1960), Low (1960), Marshall and Deshpande (1960) and Taylor (1960) point to the difficulty of applying solution chemistry thermodynamics to individual cations in relation to clay surfaces. As pointed out by Bolt (1960), Eq. [1] is accurate, in that the activity of a solute is identical whether present in suspension or in its equilibrium solution. However, when using the equilibrium solution with its standard state to characterize a suspension, we forego any knowledge about the microstructure of colloid suspensions and the relationships between the amounts of ions adsorbed and the composition of the equilibrium solution. Equal amounts of an ion adsorbed by different soil colloids do not supply the same solute activity in the equilibrium solution. On the other hand, equal solute activity in the equilibrium solutions from different soils will indicate equal solute activities for the soil suspensions.

APPLICATION OF THEORY TO SOIL TESTING

The approach to soil testing under development (Baker, 1971, 1973; Council on Soil Testing and Plant Analysis, 1974) involves measurements of the change in soil amounts of adsorbed ions which are required to make soils equal with respect to solute activity within the equilibrium solution. For the cations H, K, Ca, and Mg the activity ratios as well as the individual ionic activities in the equilibrium solution are adjusted to near optimum values for most field crops. Thus, if the solution composition is calibrated, then soil calibrations are as accurate as the theoretical considerations and analytical measurements will allow. Soil calibration data are needed because it is not theoretically possible to measure solute activities or the activities of individual ions exactly as they exist in soils under field conditions. For example, the approach with procedures published in the *Handbook on Reference Methods for Soil Testing* (Council on Soil Testing and Plant Analysis, 1974), does not measure changes in ionic activity resulting from electrostatic charge effects within the inner portion (Stern layer) of the soil solution. Results of Hornick suggested that differential exchange of K, Ca, and Mg by soils was relatively small for equilibrium solutions of chlorides, acetates, and sulfates.[3]

From previous work with nutrient solutions (Epstein & Hagan, 1952) and clay suspensions (Baker & Low, 1970; Marshall, 1960; Marshall & Upchurch, 1953), it has been concluded that the uptake of an ion by plants is determined by its relative partial molal free energy (active uptake) or the activity (passive uptake) at the root site of absorption as affected by the same parameters for other ions which interfere with the uptake of the ion being considered.

The validity of the new approach to soil testing will ultimately depend on how well the following assumptions hold or are removed by additional analytical measurements:

(a) Chloride adsorption and distribution remains relatively constant over soils when the soil to solution ratio remains constant at 1:10 (weight/volume) and the initial solution composition is:
    $50 \times 10^{-4}\,M\ CaCl_2$
    $10 \times 10^{-4}\,M\ MgCl_2$
    $2.5 \times 10^{-4}M\ KCl$
    $4 \times 10^{-4}M$ DTPA (diethylenetriaminepentaacetic acid)
    pH adjusted to 7.3 with TEA (triethanolamine).

(b) For soils tested, ionic activities are affected equally by electrostatic forces within the Stern layer around colloidal particles.

(c) Molecular interactions remain relatively constant for equilibrium testing solutions from different soils. Thus, for macro elements the activity, $\bar{a}$, is measured equally well for different soils by the concentration, C, in the testing solution.

(d) The effect that one ion, Mg, or a group of ions, K, Ca, Mg, and H, has on the uptake of another ion, K, may be measured by various Donnan ratios:

$$\frac{K}{\sqrt{Mg}}_{\text{solution}} = \frac{K}{\sqrt{Mg}}_{\text{suspension}} \quad [5]$$

and

$$\frac{K}{K + \sqrt{Ca} + \sqrt{Mg} + H}_{\text{solution}} = \frac{K}{K + \sqrt{Ca} + \sqrt{Mg} + H}_{\text{suspension.}} \quad [6]$$

[3]S. B. Hornick, 1974. Differential exchange of cations in soils. M.S. Thesis in Agronomy, The Pennsylvania State University. 81 p.

(e) For microelements, their activities in the soil suspension are reflected by, but not necessarily equal to, the values calculated from their concentrations, pH, and formation constants with DTPA after 24 hours. (R. M. Eshelman, 1975. Unpublished data. The Pennsylvania State University, Department of Agronomy).

Of the above assumptions, with computer calculations according to Debye-Huckel theory, (c) above is not necessary. However, for the electrolytes contained in the solution initially, the activity coefficients are 0.73 for $CaCl_2$ and $MgCl_2$ and 0.97 for KCl. These values approach unity compared to activity coefficients, $\bar{\gamma}$ values, of 0.018 to 0.02 obtained for NaCl in clay systems where Cl^- was held at $10^{-4}M$ and a 5% clay suspension supplied a Na^+ activity associated with near maximum uptake (Baker & Low, 1970). Marshall's discussion of Donnan relationships, Jarusov effects, and microstructures in relation to cationic bonding by clay minerals is highly recommended for an understanding of cationic activities in clay and soil systems (Marshall, 1960, 1964). From these relationships, experimental work relative to assumption (b) is most needed. For soils which adsorb Cl^-, the first assumption might lead to erroneous conclusions for macro cations. For these soils a measurement of $\bar{a}_{Cl}$ with a chloride electrode half-cell could be helpful. Currently research of graduate assistants with the Pennsylvania State University program relates to the validity of the assumptions and the application of the approach to environmental quality and implications about animal and human health from soil enrichments with chemical elements contained in sewage sludge.

SOIL TESTING PROCEDURES RELATED TO THEORY

Obtaining a solution from soils for analyses is of major importance if the results are to reflect activities and activity ratios for ions under natural conditions. For any soil testing method, the test value for an element results from an overall process or "system of causes" resulting from sampling, soil properties, analytical operations, and instrumental methods. For the new approach to soil testing (Baker, 1971), solution values for K and Mg in relation to Ca and pH were obtained using published data for the saturation percentages of the cation exchange capacity (CEC), Woodruff's energy of exchange concepts (Woodruff, 1955), and ionic activity measurements by use of membrane electrodes. Theoretical considerations discussed earlier required that certain additional conditions be met to allow a reliable estimate of ionic activities and activity ratios under field conditions from test levels in the laboratory:

(a) The sample to be tested must represent the soil to be treated with commercial fertilizer and lime. For soils and crop combinations where a substantial removal of ions occurs from horizons below the Ap, such information should be obtained by testing with depth and making these results a part of the data recorded for the soil series or productivity group.

(b) The testing solution must promote flocculation of the soil colloids to maintain maximum stability of aggregates and minimize the effect of a net electrical charge on particle surfaces on the distribution of ions in the soil solution.
(c) Trituration of aggregates must be prevented while the system is approaching equilibrium. The activity at equilibrium is a function of interparticle distances (Baker & Low, 1970).
(d) The extent to which all soils attain equilibrium should be maintained relatively constant by holding soil-solution contact constant. One hour in an end-over-end shaker followed by a 23-hour standing time has been adopted. This period of time is excessive for macro cations and deficient for attainment of equilibrium for micro elements (Baker, 1973).
(e) The net movement of solute from one phase to the other must be negligible. The concept of small exchange (Marshall, 1964) requires conditions in which:

$$\lim_{n \to 0} \frac{d\overline{G}}{dn} = 0 \qquad [7]$$

where n refers to the number of moles of solute exchanged between the suspension and dialyzate phases.

If a significant amount of solute moves, $\overline{G} - \overline{G}^\circ$ will be different from that under natural conditions. The amount of movement of a particular ion associated with a negligible change in the partial molal free energy is a function of the buffer capacity or quantity-intensity relations for the soil tested. In addition, n increases as the solute concentration of the testing solution and the solution to soil ratio increase.

Since it is not possible to meet the conditions of Eq. [7] for soils containing more or less of an element than is required for optimum activity, another more reasonable assumption would be that the activity coefficient for the solute remains constant for each soil suspension over the normal availability range:

$$\overline{a} = \overline{\gamma}\,\overline{C} \qquad [8]$$

where $\overline{a}$ is the solute activity, $\overline{\gamma}$ is the activity coefficient reflecting the bonding energy of colloids within the soil (Marshall, 1964) and $\overline{C}$ is the concentration. When used in conjunction with assumption (a) that Cl^- adsorption and distribution remains relatively constant over soils, terms of Eq. [8] become directly applicable to cationic species. From actual ionic selectivity curves, clay and soil systems show a range of constancy in cationic bonding energies; but there are large departures when the degree of charge saturation by one cation is low relative to other exchangeable cations (Marshall, 1964).

PREDICTING NUTRIENT REQUIREMENTS FROM ACTIVITY MEASUREMENTS

pH and Exchangeable Acidity

An excellent report on the subject of pH and lime requirement determination has been published recently by McLean (1973, 1975). Since soil pH and lime requirement determinations from changes in pH buffer solutions are tests which are routine to almost all soil testing laboratories, the theoretical and procedural aspects of soil testing considered within the previous sections have been applied for pH results on widely different soils in Table 1 (J. Dragun, 1975. Unpublished data. The Pennsylvania State University, Department of Agronomy). For the results reported in Table 1,

Table 1. Effect of soil to solution ratios and position of the reference electrode on pH of various testing solutions used on widely different soils (J. Dragun, 1975, Unpublished data) Pennsylvania State University, Department of Agronomy

	Solution			
	Deionized water	$SrCl_2$ 0.001 *M*	$CaCl_2$ 0.01 *M*	SMP buffer
Soil	Means over ratios and positions			
Marshall—Mo.	5.54	5.30	5.17	6.00
Guernsey—Pa.	4.93	4.57	4.28	5.38
Hagerstown—Pa.	7.59	7.50	7.13	7.08
Unclassified—Australia	5.35	5.22	5.05	5.30
Unclassified—Jamaica, W.I.	5.77	5.73	5.55	6.50
Mean	5.84	5.66	5.44	6.05
Soil:Solution (wt:vol)	Means over 5 soils			
1:1	5.65	5.62	5.42	5.49
1:2	5.77	5.67	5.41	5.94
1:10	6.10	5.70	5.48	6.72
Reference Electrode	Means over 5 soils			
UP	5.85	5.67	5.44	6.06
DOWN	5.82	5.66	5.44	6.05
F-Ratio for:				
A. Soils	1681**	33250**	6.67**	116105**
B. Soil:Solution Ratio	142**	88**	0.01	132443**
C. Ref. Electrode Position	1.41	4.3*	0.0	25.7**
A × B	11**	41.6**	0.0	6170**
A × C	0.59	1.33	0.0	1.4
B × C	0.42	0.06	0.0	8.6**
Error Mean Square	0.0038	0.00022	0.0	0.000029

*,** Significant at the 0.05 and 0.01 levels, respectively.

the samples were allowed to equilibrate for 24 hours with no mechanical agitation or dispersion of the samples. The soils were selected to represent mollisols (Marshall), two ultisols of Pennsylvania and two oxisols. The soil to solution ratios represented those in use by various laboratories (Peech, 1965). The positions of the reference electrode were altered to demonstrate the significance of the suspension effect (Marshall, 1964; McLean, 1975) under the various experimental conditions. With freshly prepared and dispersed hydrogen clay where the clay is settled out via centrifugation, pH values for the reference electrode "down" in the suspension will often be 2 pH units or more lower than when the reference electrode is left "up" within the solution phase. A small suspension effect was observed for 0.001 *M* $SrCl_2$ and SMP buffer under the conditions of the experiment, but the effect was of minor importance compared with the effect of soils and the soil to solution ratios. The results for 0.01 *M* $CaCl_2$ demonstrate the principles discussed by Peech (1965) which led to his recommendation of the use of 0.01 *M* $CaCl_2$ with the reference electrode positioned in the solution phase, "up." Not only does the use of 0.01 *M* $CaCl_2$ reduce the effect of changes in the soil to solution ratio, it also decreases the effect of differences in the initial electrolyte concentration as well as other effects leading to a statistically significant interaction involving soils and soil to solution ratios. The pH of soil in 0.01 *M* $CaCl_2$ is usually about 0.5 pH unit lower than that measured in water. Averaged over all soil to solution ratios, the difference between pH measured in water and in 0.01 *M* $CaCl_2$ varied from 0.22 to 0.65 unit. Which pH measurement best represents the acidity of the soils under field conditions? The theoretical considerations with respect to Eq. [1] to [4) would favor either the use of the 0.001 *M* $SrCl_2$ as developed by Goedert and Corey (W. J. Goedert, and R. B. Corey, 1973. Availability of indices for soil cations by a single equilibration with $SrCl_2$. Agron. Abstr. 73-82) or the use of 0.01 *M* $CaCl_2$ as initially proposed by Schofield and Taylor (1955) as discussed by Peech (1965), McLean (1973, 1975) and Woodruff (1967). For either chloride salt, it must be assumed that:

(a) The amount of soluble electrolytes contributed by the soil is insignificant compared to the amount of salt added. Therefore, as the chloride concentration decreases, the volume of solution per unit of soil must increase.

(b) The concentration of metal chlorides in solution is approximately equivalent to the total electrolyte concentration of the soil solution under actual field conditions. From the development of a new approach to soil testing (Baker, 1973; Council on Soil Testing and Plant Analysis, 1974), the electrolyte concentration of a soil under field conditions was calculated to be equivalent to approximately:

50×10^{-4} *M* $CaCl_2$
10×10^{-4} *M* $MgCl_2$
2.5×10^{-4} *M* KCl

Thus, if a measured pH is desired to predict the pH under field conditions, the concentration of $CaCl_2$ should be approximately 61.25×10^{-4} *M* or 0.006 *M*. We could conclude then that the mean pH of the soils of Table 1 under field conditions is less than 5.66, the value for 0.001 *M* $SrCl_2$ and near 5.44, the value for 0.01 *M* $CaCl_2$. Since pH is not of major biological significance except as it reflects the solubilities and availabilities of several trace elements and phosphorus, the use of 0.01 *M* $CaCl_2$ at a 1:1 soil to solution ratio seems appropriate for soils of medium to fine texture. For soils of low CEC, the measured pH could be in error because the conditions stipulated for Eq. [7] and [8] are not met. The capacity for exchange among cations is 2 meq/100g for 0.01 *M* $CaCl_2$ at a soil to solution ratio of 1:1.

(c) As indicated by Peech (1965), the use of 0.01 *M* $CaCl_2$ will keep the soil suspensions flocculated. Not only does flocculation decrease the magnitude of the errors due to the suspension effect, it also prevents a change in ionic activity associated with changes in interparticle distances or dispersion.

Until more is known about the activities of ions in different soils in relation to their biological activities, the use of pH in 0.01 *M* $CaCl_2$ at a 1:1 soil to solution ratio using the procedures outlined recently by McLean (1975) could make comparisons of results among laboratories and soils much more meaningful.

With respect to exchangeable acidity or lime requirement, we have in the buffer pH measurement an example of how ionic activity data may be used to determine directly the corrective treatment for a soil (McLean et al., 1966). If the amount of soil acidity reflected by a given change in buffer pH is constant, then it should be possible to change the soil to solution ratio and still obtain a constant lime requirement. When this was done for the results for the SMP buffer (Table 1) the average lime requirement over all soils would be 23.3, 32.3 and 40.3 metric tons/ha (10.4, 14.4, and 18 tons/acre) for the 1:1, 1:2, and 1:10 ratios, respectively. Thus it is not possible to obtain a constant lime requirement by the SMP buffer from changes in pH when the soil to solution ratio is altered. For a buffer to be theoretically accurate for measuring the lime requirement, the following conditions should be met:

(a) The initial pH of the buffer should be equal to the pH desired at the surface of plant roots under field conditions.
(b) The buffer solution must be capable of reaching a reversible equilibrium with all forms of soil acidity.
(c) The change in total acidity with a change in pH of the buffer must be known and be approximately equal to that of the soils tested.

With respect to factors (a) and (c) above, the Woodruff buffer method (Woodruff, 1948) has considerable advantage over more recently developed methods (Adams & Evans, 1962; Shoemaker et al., 1962). The capacity of the Woodruff buffer at the recommended soil to solution ratio is 1.0 meq acidity/100g soil or 1,120 kg/ha of $CaCO_3$ required for each tenth unit decrease in pH. For medium textured soils with much pH dependent CEC, the buffer capacities, $\Delta Q_H/\Delta pH$, are relatively constant and equal to that of the Woodruff buffer when used at a soil to solution ratio of 1:2. From theoretical changes in relative partial molal free energies of HCl with changes in pH as developed in Eq. [3] and illustrated in Figure 1, it can be predicted that the lime requirement of a soil should be double the amounts of acidity absorbed from the soil by the buffer solution. Thus, the lime requirement determined by the Woodruff method (Woodruff, 1948) should be 2,240 kg/ha of $CaCO_3$ for each tenth unit decrease in pH for many soils. This observation seems important in light of the experimental data showing that the actual lime requirements of soils were double the amounts

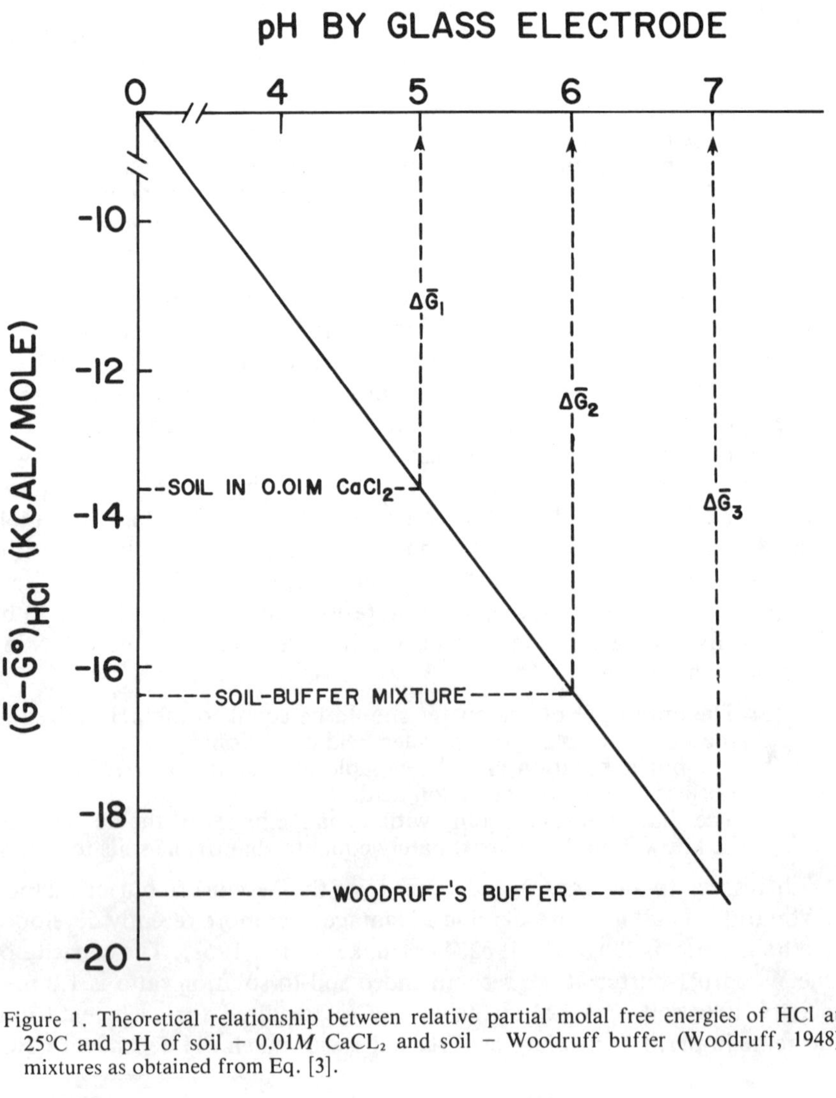

Figure 1. Theoretical relationship between relative partial molal free energies of HCl at 25°C and pH of soil – 0.01M $CaCL_2$ and soil – Woodruff buffer (Woodruff, 1948) mixtures as obtained from Eq. [3].

predicted by the Woodruff method. For the soil illustrated in Figure 1, the buffer capacity of the soil is equal to that of the Woodruff buffer. For this condition, the pH increase of an acid soil would be equal to the pH decrease of the buffer. Likewise, the decrease in free energy represented by $(\Delta\overline{G}_2 - \Delta\overline{G}_1)$ would be equal to the increase in free energy represented by $(\Delta G_2 - \Delta G_3)$. As the buffer capacity of the soil decreases, $\Delta\overline{G}_2$ will approach $\Delta\overline{G}_3$; and as the buffer capacity increases, $\Delta\overline{G}_2$ will approach $\Delta\overline{G}_1$. The amount of acidity, ΔQ_H, neutralized by the buffer does not include the amounts remaining from the equilibrium pH to pH 7.0. By adjusting the buffer concentration or the soil/buffer ratio to obtain equal buffer capacities for soils and the buffer, the total exchangeable acidity would be the amounts absorbed by the buffer. The Woodruff buffer (Woodruff, 1948) has been criticized with respect to condition (b), previously mentioned, and it has been postulated that it does not react with soil acidity associated with high concentrations of exchangeable Al in soils (McLean, 1975; Shoemaker et al., 1962). Theoretically, the Woodruff buffer could be used equally well for soils of high and low buffer capacities. For soils low in buffer capacity (CEC of approximately 5 meq/100g), the Woodruff buffer could be used at half the normal concentration and the lime requirement should actually be 1,120 kg/ha of $CaCO_3$ for each 0.1 unit decrease in pH.

A New Approach to Soil Testing

The conditions stated above with respect to the determination of lime requirements of soil from buffer pH may be applied for predicting soil requirements from activity measurements of other cations where the activity of chloride remains constant over soils. Details of the testing procedure have been included in the *Handbook on Reference Methods for Soil Testing* (Council on Soil Testing and Plant Analysis, 1974). The interpretation of the test values in relation to corrective treatments of soils are of major concern here. As was shown for the determination of lime requirement using buffer capacities and buffer pH measurements, corrective treatments are more easily and accurately measured than are the initial activities under field conditions. Should soil test values or numbers calculated from them be correlated significantly with plant responses? As biologists, we say yes; but, from the point of view of the grower, the correlations are considered academic. Only by assuming a constant chloride activity and equal quantity-intensity relations for all soils tested is it possible to calculate an initial solute activity, $\bar{a}$, existing under field conditions using the new approach to soil testing. From the initial concentration of an ion in the testing solution, C_i and the change in concentration, ΔC_i, resulting from equilibration of the testing solution with the soil,

$$\bar{a}_i = C_i + N\Delta C_i \qquad [9]$$

where N is a function of the buffer capacity of the soil in relation to the soil to testing solution ratio. For soils differing substantially with respect to CEC and/or mean free bonding energies for an ion, i, (Marshall, 1960, 1964), one should not expect a high correlation between soil test results and plant responses. The approximate validity of Eq. [9] has been demonstrated by Stout from results for two ultisols which differ substantially with respect to fixed K.[4] From a biological point of view as well as from a theoretical point of view, Eq. [9] should be used only to obtain values for use in Eq. [5] and [6] to characterize soil with respect to selectivity ratios.

The conditions stated earlier for calculating $\bar{a}$ from soil test values by the new method (Council on Soil Testing and Plant Analysis, 1974) are not necessary for determining corrective treatments. With respect to most elements essential for plants, soil testing can provide for *efficient* use of fertilizers with no obvious crop response to an additional increment of a particular element. The needs and objectives of soil testing are made clear from a study of Figure 2 as modified from Luckey (1975). For many elements essential for plant growth, there is a wide range of tolerance from the activity which is just sufficient for maximum yields and that which is toxic. For example, results of Dragun showed that corn (*Zea mays* L.) plants tolerate solution activities of Cu ranging from 10^{-13} to 10^{-6} M.[5] While the uptake of ions and yield responses for several ions follow the relationship shown in Figure 2, the associated activity levels are different for the various elements. For nonessential, toxic elements, negative responses are more often associated with animals consuming the plants than by the plants themselves. A discussion of soil testing in relation to toxic elements has been presented by Baker and Chesnin (1975).

From results published previously (Baker, 1971, 1973) and the results of Stout[6] which considered K and Mg as analogous to H of Figure 1, the K and Mg requirements may be determined directly from the following equations using testing solution A described in the *Handbook on Reference Methods for Soil Testing* (Council on Soil Testing and Plant Analysis, 1974):

$$\text{K (requirement in ppm)} = -(78 \times 10^4)\Delta\text{K} \qquad [10]$$

$$\text{Mg (requirement in ppm)} = -(48 \times 10^4)\Delta\text{Mg} \qquad [11]$$

where ΔK and ΔMg are equal to ΔC_i of Eq. [9] for the respective ions. Positive ΔK and ΔMg values lead to negative soil requirements which represent the amounts in soil greater than those required for minimum levels of availability. In the original work (Baker, 1971), amounts of each element required to return the solution to its original concentration were

[4] W. L. Stout, 1975. Unpublished data. The Pennsylvania State University, Department of Agronomy.

[5] J. Dragun, 1975. Substrate copper activities and uptake by corn hybrids. M.S. Thesis in Agronomy, The Pennsylvania State University. 86 p.

[6] Stout, 1975. Unpublished data.

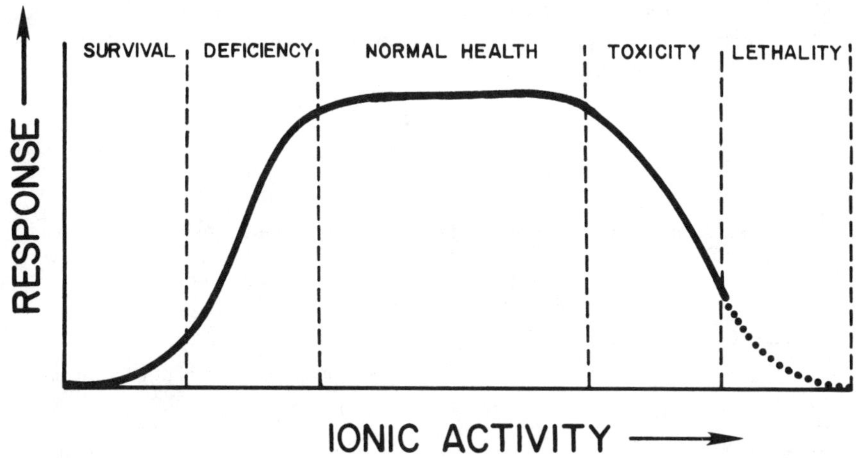

Figure 2. Substrate ionic activity-response relationship of an essential element as modified from Luckey (1975).

not considered. Therefore, the constants shown for Eq. [10] and [11] are double those reported previously (Baker, 1971). For example, how much Mg would be required to provide a Mg activity of 10×10^{-4} molal at plant root surfaces, if the level in the testing solution decreased from 10×10^{-4} to 6×10^{-4} molal? At the soil to solution ratio of 1:10, 100g of soil would be associated with a liter of solution, so it would have adsorbed 4×10^{-4} moles of Mg. This amount of Mg based upon the weight of the soil would be equivalent to 0.8 meq/100g or 96 ppm. However, to return the solution to its original Mg concentration would require another 96 ppm or a total Mg treatment of 192 ppm.

Results of Hornick[7] indicated that DTPA in the soil testing solution did not affect the equilibrium concentrations of Ca, Mg, and K substantially when the DTPA concentration did not exceed 4×10^{-4} *M*. However, Ca requirements as lime should be determined from buffer pH rather than from changes in Ca in the testing solution. Test levels for Ca of less than 45×10^{-4} *M* have been associated with high levels of exchangeable H, NH_4, Na, Mg, K, or combinations of cations in different soils. Test levels for Ca of 60 to 75×10^{-4} *M* have been associated with free $CaCO_3$ or high soluble salt concentrations. High Ca from free $CaCO_3$ will be reflected by an equilibrium pH greater than the initial testing solution pH of 7.3. If soluble salts are high, the corrective treatments calculated for K and Mg will be low. Some compensation for variable concentrations of soluble salts could be accomplished by adjusting the equilibrium concentrations of K and Mg relative to the original salt concentration in the testing solution. The adjustment may be accomplished by multiplying the respective test concentrations by the ratio of $122.5 \times 10^{-4}/\Sigma$ Ca + Mg + K + Na. For

[7]Hornick, 1974. Differential exchange of cations.

the corrections, cations are expressed in equivalents per liter. Compensation in this manner assumes small exchanges resulting from electrolytes present in the soil initially.

Table 2. Yields of corn grain and soil and plant analysis results for field plots on Rayne channery loam as affected by five annual rates of plowdown P with and without row fertilizer to supply 34 to 56 kg/ha of P_2O_5 from 10-20-10 or 8-24-8 fertilizer. Soil test and plant analysis results for 1973

	Grain yields			Soil test P		Plant composition		
P_2O_5	1971	1972	1973	Bray 1	New	P	Cu	Zn
kg/ha					μm/liter	%	ppm	
0	4959	4143	5021	31	0.29	0.35	22	57
84	5523	4331	5120	46	0.49	0.38	19	55
168	5837	4394	5649	57	0.67	0.44	19	53
336	6402	4582	5147	97	0.40	0.45	19	58
672	6590	4394	5523	214	0.70	0.47	16	52
Significance	*	N.S.	N.S.	**	N.S.	**	*	N.S.
– Banded	5460	3892	5210					
+ Banded	6214	4833	5398					
Significance	**	**	N.S.					

*,** indicate significant differences at the 0.05 and 0.01 levels of probability, respectively. Correlation coefficients (r) relating: Bray 1 vs. New Test = 0.51*; Bray 1 vs. Plant P = 0.76**; New Test vs. Plant P = 0.44.

Table 3. Yields of corn grain and soil and plant analysis results for field plots on Rayne channery loam as affected by rates of K. Soil test and plant analysis results for 1973. Correlation coefficients (r)† involving K by the new method were derived from activities calculated from Eq. [9] assuming N = 2

K_2O	1971	1972	1973	Soil test K exchangeable		New test K	Plant K	Plant Mg
kg/ha	kg/ha			meq/100 g	% sat.	M/liter × 10^4	%	
0	5586	4708	4833	0.18	1.47	1.39	1.46	0.46
84	5586	3584	5021	0.20	1.63	1.54	1.88	0.29
168	6151	4519	5335	0.32	2.63	2.03	2.17	0.20
336	5837	4205	5398	0.51	3.93	2.84	2.21	0.17
672	5962	4519	5461	0.83	6.20	4.25	2.11	0.16
Significance	N.S.	N.S.	*	**	**	**	**	**
– Banded	5335	3766	5084					
+ Banded	6276	4959	5335					
Significance	**	**	N.S.					

*,** indicate significant differences at the 0.05 and 0.01 levels, respectively.

† Correlation coefficients (r) relating: K meq/100 g vs. K % Sat. = 0.98**; K % Sat. vs. New Test = 0.996**; K New Test vs. Plant K = 0.51*; K % Sat. vs. Plant K = 0.51*; Plant K vs. Plant Mg = –0.87**.

Correlations with Plant Analysis

Corn grain yields, soil test results, and plant composition data for experiments on Rayne channery loam are shown in Tables 2 to 5. The field

Table 4. Yields of corn grain and soil and plant analysis results for field plots on Rayne channery loam as affected by rates of calcitic and dolomitic limestone added in 1971. Soil test and plant analysis results for 1973

Lime treatments		Grain yields			Soil test results				Plant
Total	Mg	1971	1972	1973	pH	Ca	Mg	New Mg	Mg
kg/ha						% sat.		M/liter × 10^4	%
1,345	84	5586	3982	4645	4.7	25	4.0	9.6	0.14
11,200	0	5398	4582	5147	5.4	52	2.6	8.1	0.10
11,200	84	5335	4331	5837	5.4	53	3.1	8.5	0.12
11,200	168	5586	4456	5461	5.7	59	4.0	8.8	0.13
11,200	336	5649	4582	5271	5.5	51	5.9	9.7	0.14
5,600	336	6026	5331	5398	5.0	31	6.0	10.3	0.19
2,800	336	5398	3452	4582	4.5	17	4.8	10.2	0.20
22,400	336	5712	4394	5398	6.1	59	3.1	8.4	0.14
11,200	672	5210	4431	5335	5.3	40	9.1	11.3	0.21
Significance		N.S.	**	*	**	**	**	**	**
− Banded		5021	3892	5272					
+ Banded		6088	4645	5210					
Significance		**	**	N.S.					

*,** indicate significant differences at the 0.05 and 0.01 levels, respectively. Correlation coefficient (r) relating: pH to Ca % Sat. = 0.88**; New Test Mg to Plant Mg = 0.84**; Mg % Sat. to New Test = 0.88**; Mg % Sat. to Plant Mg = 0.70*.

Table 5. Results for new soil test and corn leaf analysis for Fe, Mn, Cu, Zn, and Al as affected by rates of calcitic and dolomitic lime applied in 1971 to supply variable rates of Mg. Soil test and plant analysis results for 1973

Lime treatments		Soil	Plant	Soil	Plant	Soil	Plant	Soil	Plant	Soil	Plant
Total	Mg	Mn	Mn	Fe	Fe	Zn	Zn	Cu	Cu	Al	Al
kg/ha		ppm									
2,800	84	4.9	628	1.4	261	0.18	61.3	0.09	18.0	6.0	102
11,200	0	1.9	201	1.4	286	0.09	47.7	0.09	18.7	3.2	87
11,200	84	2.0	208	1.2	218	0.07	54.7	0.09	17.7	3.0	67
11,200	168	1.4	176	1.2	249	0.08	50.7	0.10	15.3	2.5	105
11,200	336	1.6	150	1.3	196	0.08	50.7	0.09	16.7	3.0	71
5,600	336	3.3	331	1.4	249	0.09	58.0	0.08	19.0	5.1	81
2,800	336	4.9	482	1.4	257	0.13	57.0	0.07	18.7	6.3	89
22,400	336	0.9	107	0.9	223	0.07	40.3	0.09	16.0	1.6	80
11,200	672	2.4	169	1.3	227	0.08	55.7	0.09	18.0	3.8	78
		**	**	N.S.	N.S.	**	*	*	*	**	N.S.

*,** indicate significant differences at the 0.05 and 0.01 levels of probability, respectively. Correlation coefficients (r) relating: New Test Mn vs. Plant Mn = 0.91**; New Test Fe vs. Plant Fe = N.S.; New Test Zn vs. Plant Zn = 0.52*; New Test Cu vs. Plant Cu = N.S.; New Test Al vs. Plant Al = N.S.

location was chosen on the basis of results by Bair which indicated the soil was low in pH, P, K, Ca, Mg, and Zn (N. J. Bair, 1971. Development of diagnostic procedures to determine the reasons for poor crop growth. M.S. Thesis in Agronomy, The Pennsylvania State University. 72 p.). Further investigation revealed that the field had not been cultivated for at least 40 years. The objective of the experiments was to establish soil test correlations for existing soil tests and the new approach to soil testing with the management practice of banded fertilizer near the row at planting. When soil fertility levels of the selected site are considered relative to Eq. [9], Figure 2, and the concept of soil testing as a management tool to remove limiting factors for crop production, the experiments could be considered as examples of obtaining correlation data for past soil fertility programs. The author would agree with that conclusion. On the other hand, only field results could be used to determine the test levels associated with minimum activities for normal plant growth. In these experiments only plant responses to macro elements and pH changes were included.

The pH (1:1 in water) was 5.0 with a soil lime requirement by the Woodruff method (1948) of 11,200 kg/ha of $CaCO_3$. The test level for P by the Bray 1 method (Bray & Kurtz, 1945) was 13 kg/ha, and the K and Mg saturation of the cation exchange capacity (CEC of 12.3 meq/100g) was 1.3% and 1.0%, respectively. When not included at variable rates in one of the experiments, the plots were fertilized each year to supply 224 kg/ha of N, 336 kg/ha of P_2O_5, and 336 kg/ha of K_2O. In 1971 only, 336 kg/ha of Mg from dolomitic limestone and a total limestone treatment of 11,200 kg/ha were applied to the P and K experiments.

The P experiment consisted of 0, 84, 168, 336, and 672 kg/ha of P_2O_5 applied in the spring each year and plowed under; the K experiment consisted of the same rates used for P_2O_5; and the variable lime and Mg treatments of the liming experiment are defined in Table 5. The plots of each experiment were arranged as randomized blocks of three replications. Each plot was split to give 4 rows of corn receiving banded fertilizer and 4 rows without banded fertilizer. Only the two center rows of each split were sampled for leaf analysis and harvested for yield.

Under the conditions of the P experiment, the response to banded fertilizer was substantial during the first 2 years. While the use of banded fertilizer reduced the correlations between soil test levels and plant content of P, the results illustrate vividly the importance of banded P in these soils. It is interesting to note that after a total treatment of 2018 kg/ha of P_2O_5, the level of soluble P had attained a concentration of only 0.7 μm/liter. The correlation of 0.76 relating plant P to soil test P by the Bray 1 method was superior to the coefficient of 0.44 relating plant P to soil test P by the new method. As currently developed, the Bray 1 method is superior on these soils for determining the soil P requirements (Baker & Hall, 1967), while the new method is valuable for predicting a response to banded P. However, within a soil group, the Bray 1 method with calibration data is considered to be of more value than the new method. For soils of Pennsylvania, a test level of 67 kg/ha of P by the Bray 1 method with band placement of P for cereal grains is considered adequate. If this level of P by the Bray 1 method were associated with test levels by the new method of 2 μm/liter or higher, no response to banded P would be expected.

For the K experiment, Table 3, the response to banded fertilizer was considered primarily as a response to P. The results illustrate several relationships important to soil testing. From previous work, the response to added K has been low when the K saturation exceeded 2%. From theoretical considerations (Baker, 1971), it was calculated that for the average soil a 2% saturation of the CEC should be equivalent to a KCl activity of 2.5×10^{-4} *M* by the new method. For this soil the bonding energy for exchangeable K is higher than average because the KCl concentration in solution of only 2.03×10^{-4} *M* was associated with a K saturation of 2.63%. Nevertheless, the correlation coefficient of 0.996 indicates the close association between the two methods where the composition of the solid phase of the soil is constant. The correlation coefficient of 0.51 relating soil test levels to plant K was equal by both methods. The highly significant negative correlation, −0.87, illustrates the substantial effect of increased soil K activity on decreased Mg uptake (Baker, 1972). If 2% K in corn leaves is taken as a critical level for K in plants, then the KCl activity representing a minimum activity within the soil could be reduced from 2.5×10^{-4} *M* to 2.0×10^{-4} *M*. As shown previously, this would amount to a change in exchangeable K of about 20 ppm within the soil. Since crop removal of K is relatively great, most soils at the critical level, represented by a KCl concentration in solution by the new method of 2.0 to 2.5×10^{-4} *M*, will not supply adequate amounts of K over the growing season for some crops. Management of these crops should include soil K applications above the amounts required for the critical soil test level to meet the crop requirements. High K soils release the amounts of K required for many crops giving a zero requirement.

The results for the liming experiment would have been more valuable if the lime treatments had been higher, Table 4. The highest rate of lime, which was double the requirement as determined by the Woodruff method, was not sufficient to bring the pH (1:1 in water) to 6.5 or above after 3 years. As indicated previously, the lime requirement by the Woodruff method was theoretically and actually near 22,400 kg/ha instead of 11,200 kg/ha. While the correlation coefficient relating Mg saturation of the CEC to Mg by the new method was high (r = 0.88), the plant Mg was better correlated with Mg by the new method (r = 0.84) than by the Mg saturation (r = 0.70). As indicated in the previous report (Baker, 1971), a Mg saturation of 6% was closely related to a $MgCl_2$ concentration in solution of 10×10^{-4} *M*. At the rates of K applied (336 kg/ha of K_2O/year) only the highest rate of Mg (672 kg/ha in 1971) with 11,200 kg/ha of lime, or 336 kg/ha Mg at 2,800 and 5,600 kg/ha of lime, were associated with test levels by the new method of 10×10^{-4} *M* and plant concentrations of 0.2% which is the critical level for Mg in relation to hypomagnesemia in cattle (Baker, 1972). These results illustrate the im-importance of such ionic ratios as $K/\sqrt{Mg}$, Ca/Mg, and $H/\sqrt{Ca + Mg}$ within soils in relation to plant composition (Baker, 1972).

For microelements, the new approach to soil testing shows much promise. The results presented in Table 5 indicate a high correlation relating plant Mn to soil test Mn, and the effect of lime treatments on soil test levels. Results by Dragun[8] indicate that analysis of corn roots would improve correlations between soil test levels and plant response to Cu and other less mobile microelements. From experience gained using DTPA (Viets & Lindsay, 1973) and solution A as described in the *Handbook on Reference Methods for Soil Testing* (Council on Soil Testing and Plant Analysis, 1974), several microelements may be tested for simultaneously. Corrective treatments of soils and the use of soils for waste disposal must maintain levels adequate for efficient crop production without attaining levels harmful to the food chain as indicated by Figure 2.

Ionic ratios have not been stressed in the presentation of data, but the results clearly indicate the importance of such ratios as $K/\sqrt{Mg}$, Ca/Mg, etc. These ratios have been discussed previously (Baker, 1972). For the new approach to soil testing, the chemical composition of the testing solution was designed to provide corrective treatments to maintain minimum levels of activity and optimum ratios within soils.

In summary, it has been established that an approach to soil testing can be developed for all essential and nonessential elements which is somewhat analogous to the Woodruff test for pH and lime requirement. Needless to say, the problems for each element are also analogous to those encountered with pH and lime requirement testing. Should the soil pH be adjusted to 6.0, 6.5 or 7.0 for a particular crop? The optimum pH is a function of microelement availability, Al solubility, and crop (Baker, 1972). Should the K level of the soil be adjusted to provide a KCl concentration of 2.0 to 2.5×10^{-4} molal with differential fertilization to meet the requirements of different crops or should the level be adjusted to higher values to meet the K requirements of all crops? The later choice will lead to an increased incidence of hypomagnesemia in cattle which will require heavy applications of Mg with no crop yield response. The theoretical considerations and the results to date using solution A as described in some detail in the *Handbook on Reference Methods for Soil Testing* enable the following conclusions:

(a) Ionic activities and ratios within soil are of major biological significance in crop production.
(b) Soil testing methods analogous in theory and application to existing tests for pH and lime requirement using buffer solutions are possible for many adsorbed elements important in plant nutrition and animal and human health.
(c) An adjustment of ionic activities to minimum levels in all soils as required for optimum nutrition of field crops is recommended.

---

[8]Dragun, 1975. Substrate copper activities and uptake.

(d) Corrective treatments of soils to provide minimum ionic activities should also maintain optimum ionic activity ratios for both crop production and animal health.
(e) Crop fertilization should be managed separately from corrective treatments of soils, especially for elements such as N and K where uptake is high compared with the amounts required within soils for minimum activites.
(f) For trace elements, soil activities should be measured and maintained within limits which enable efficient crop production without harmful effects on the food chain.

LITERATURE CITED

Adams, F., and C. E. Evans. 1962. A rapid method for measuring lime requirement of red-yellow podzolic soils. Soil Sci. Soc. Am. Proc. 26:355-357.

Baker, D. E. 1971. A new approach to soil testing. Soil Sci. 112:381-391.

Baker, D.E. 1973. A new approach to soil testing: II. Ionic equilibria involving H, K, Ca, Mg, Mn, Fe, Cu, Zn, Na, P, and S. Soil Sci. Am. Proc. 37:537-541.

Baker, D. E. 1972. Soil chemistry of magnesium. p. 1-39. *In* Magnesium in the Environment. Taylor County Printing Co., Reynolds, Ga.

Baker, D. E., and L. Chesnin. 1975. Chemical monitoring of soils for Environmental Quality and Animal and Human Health. Adv. Agron. 27:305-374.

Baker, D. E., and J. K. Hall. 1967. Measurements of phosphorus availability in acid soils of Pennsylvania. Soil Sci. Soc. Am. Proc. 31:662-667.

Baker, D. E., and P. F. Low. 1970. Effect of the sol-gel transformation in clay-water systems on biological activity: 2. Sodium uptake by corn seedlings. Soil Sci. Soc. Am. Proc. 34:49-56.

Baker, D. E., and C. M. Woodruff. 1963. Influence of volume of soil per plant upon growth and uptake of phosphorus by corn from soils treated with different amounts of phosphorus. Soil Sci. 94:409-412.

Bolt, G. H. 1960. Cations in relation to Clay Surfaces. Int. Congr. Soil Sci. Trans. 7th (Madison, Wis.) II: 321-327.

Bray, R. H., and L. T. Kurtz. 1945. Determination of total, organic and available forms of phosphorus in soil. Soil Sci. 59:39-45.

Council on Soil Testing and Plant Analysis. 1974. Handbook on Reference Methods for Soil Testing. 2400 College Station Road, Athens, Georgia. 101 p.

Epstein, E., and C. E. Hagen. 1952. Kinetic study of the absorption of alkali cations by barley roots. Plant Physiol. 27:407-474.

Low, P.F. 1950. Force fields and chemical equilibrium in heterogeneous systems with special reference to soils. Soil Sci. 71:409-418.

Low, P. F. 1954. Ionic activity measurements in heterogenous systems. Soil Sci. 77:29-41.

Low, P. F. 1960. Influence of adsorbed water on electrochemistry of clays. Int. Congr. Soil Sci. Trans. 7th (Madison, Wis.) II: 328-336.

Luckey, T. D. 1975. Hormology with inorganic compounds. p. 81-118. *In* T. D. Luckey, B. Venugopal, and D. Hutcheson (ed.). Heavy metal toxicity, safety and hormology. Academic Press, New York.

Marshall, C. E. 1960. Colloid chemistry of soil in relation to plant nutrition. p. 723-736. *In* M. X. Zarrow (ed.) Growth in living systems. Basic Books, Inc., New York.

Marshall, C. E. 1964. The physical chemistry and mineralogy of soils. John Wiley. 244 p.

Marshall, C. E., and K. B. Deshpande. 1960. The validity of electrometric methods as applied to clay-water systems. Int. Congr. Soil Sci. Trans. 7th (Madison, Wis.) II: 337-341.

Marshall, C. E., and W. J. Upchurch. 1953. Free energy in cation interchange as illustrated by plant root—substrate relationships. J. Phys. Chem. 57:618-621.

McLean, E. O. 1973. Testing soils for pH and lime requirement. p. 78-95. *In* L. M. Walsh, and J. D. Beaton (ed.). Soil testing and plant analysis, Revised Edition. Soil Sci. Soc. Am., Madison, Wis.

McLean, E. O. 1975. Recommended pH and lime requirement tests. NCR Pub. 221, N. Dak. Agric. Exp. Stn. Bull. 499. p. 6-9.

McLean, E. O., S. W. Dumford, and F. Cononel. 1966. A comparison on several methods for determining lime requirement of soils. Soil Sci. Soc. Am. Proc. 30:26-30.

Peech, M. 1965. Hydrogen-ion activity. *In* C. A. Black (ed.), Methods of soil analysis, Part II, Chemical and microbiological properties. Agronomy 9: 914-926. Am. Soc. Agron., Madison, Wis.

Pitzer, K. S., and L. Brewer. 1961. Thermodynamics. Second edition. Revised from Lewis, G.N., and M. Randall. McGraw-Hill, New York. 723 p.

Schofield, R. K., and A. W. Taylor. 1955. The measurement of soil pH. Soil Sci. Soc. Am. Proc. 19:164-167.

Shoemaker, H. E., E. O. McLean, and P. F. Pratt. 1962. Buffer methods for determination of lime requirement of soils with appreciable amounts of exchangeable aluminum. Soil Sci. Soc. Am. Proc. 25:274-277.

Taylor, A. W. 1960. The significance of ratios of ion activities. Int. Congr. Soil Sci. Trans. 7th (Madison, Wis.) II: 342-346.

Viets, Jr., F. G., and W. L. Lindsay, 1973. Testing soils for zinc, copper, manganese and iron. p. 153-172. *In* L. M. Walsh, and J. D. Beaton (ed.). Soil testing and plant analysis. Soil Sci. Soc. Am., Madison, Wis.

Woodruff, C. M. 1948. Testing soils for lime requirement by means of a buffered solution and the glass electrode. Soil Sci. 66:53-63.

Woodruff, C. M. 1955. The energies of replacement of calcium by potassium in soils. Soil Sci. Soc. Am. Proc. 19:167-170.

Woodruff, C. M. 1967. Crop response to lime in the Midwestern United States. *In* R. W. Pearson, and F. Adams (ed.). Soil Acidity and Liming. Agronomy 12:207-231. Am. Soc. of Agron., Madison, Wis.

Nutrient Intensity and Balance[1]

C. M. Geraldson[2]

Periods of true nutrient balance are transitory at best. Nevertheless, we are obligated to study and determine combinations of controllable factors (nutrient quantities, ratios, sources, placement, and movement) that can be applied to soils which will shift the soil root environment toward an optimum nutrient intensity and balance. When, by integrating selected components, an optimum combination of components is integrated with a soil, yields can be increased toward an economic maximum. The production or yield levels vary with the system of production which from an economic standpoint is based on obtaining maximum yield and quality as efficiently as possible (Brooke, 1975).

Agronomic crop production is considered as an extensive system of production in contrast to horticulture crops which for the most part would be grown more intensively. Soils in varying degree provide some nutrients for growing crops. Soil testing has been used to estimate the quantity of nutrients available which can then be supplemented with fertilizers to provide the approximate nutrient requirement of any given crop. The validity of soil testing is based on the relevance of correlations provided by research with emphasis on the associated crop production.

In Florida, a major portion of conventional soil tests for available nutrients lack validity because the necessary correlations have been inadequate. The sandy soils of peninsular Florida provide minimal nutrients in their natural state; thus most of the nutrients for the intensively grown crops are provided by fertilizers or other supplements. Furthermore, rains during the growing season can alter the nutrient concentration and balance established by the applied combinations of controllable factors (Geraldson, 1962, 1963, 1967, 1970). Thus the dominant factors associated with crop production nutrition are the application of relatively large quantities of nutrients and the alterations that develop due to water movement and crop utilization. With the intensity and balance procedure the concept as well as the method of soil testing was altered because conventional concepts and

---

[1]Florida Agric. Exp. Stn. Journal Series No. 6062.

[2]Professor, soils chemist, IFAS, University of Florida, Agricultural Research and Education Center, Bradenton, Florida.

methods were not satisfactory. The 'intensity and balance' method of soil testing was initiated and evaluated on this basis and emerged as a system with a favorable degree of validity (Geraldson, 1955, 1957a, b, 1962, 1963, 1967, 1970).

PROCEDURE

To minimize dilution effect, soil soluble salts should be extracted from soil at moisture contents approaching field levels but also at a content that can be conveniently manipulated. The saturation extract (Richards et al., 1954) was chosen for this reason and is obtained by vacuum filtration of a soil paste that has been made up to a saturated condition by adding water while stirring. About 10cc of the saturation extract is obtained from a soil volume of 75cc. The electrical resistance of this extract is determined with a conductivity bridge. Electrical conductivity (EC), which is the reciprocal of the electrical resistance, varies directly with soluble salt concentration. EC readings can be used as an indicator of soluble salts (Richards et al., 1954). However, for interpretive reasons which will be indicated later the EC was converted to ppm salt by using KCl as a standard.

Nutrient concentration and balance in the soil solution calculated for a field moisture level was utilized for interpretation of data from analyses of the saturation extract (Richards et al., 1954). Total salt concentrations and proportionate amounts of specific ions contained in the soil solution were correlated with crop responses, and interpretations have been established (Geraldson, 1955, 1957a, 1962). Thus, with a single water extract we have the capability to evaluate the ionic environment in the soil solution/total nutrient concentrations and balances as well as salinity and pH.

The salts contained in this water extract represent an equilibrium mixture from ionic exchange systems and salts of varying solubilities diluted to a soil moisture level of saturation, which is interpolated by a calculated factor to the field moisture level. The ions are subject to changes and equilibration based on the myriad physiochemical and biological processes in the soil. However, the more drastic ionic changes in the root environment are due to addition of fertilizers and other soil supplements, movement by irrigation water and rain, and removal by crop utilization and leaching. Thus, by using the soil solution as a common denominator, similar ionic variations in soil solutions for any soil type regardless of exchange capacity, organic matter content, or water holding capacity, are essentially equivalent with respect to plant response. Considering that the soil solution is characterized by heterogeneity in micro-chemical space and the hydroponic culture represents homogeneity, then similar correlations may be obtained with ionic variations, regardless of whether they are contained in the root environment portion of the soil solution or in a hydroponic solution. As with any soil test, when sampling a heterogenous soil solution, interpretation is relevant to the portion of the soil sampled.

Therefore, a series of cores across a bed reflects an average; whereas separate increments of the soil bed profile better reflect the heterogenous nature of the soil, portions of which can be considered as the relevant root environment.

In this paper the ionic root environment will be defined and evaluated in terms of ionic concentrations and balances in the soil solution. These measures can be correlated with crop response and collectively used to describe the ionic root environment and can be altered to approach an optimum intensity and balance.

NUTRIENT INTENSITY

Nutrient intensity could be defined as a nutrient concentration or range of concentrations that is neither deficient or excessive but is an approach to an optimum for intensive production. Salt concentration (conductivity of the saturation extract) was used (Geraldson, 1955, 1957a) as a measure of nutrient intensity in the soil solution.

Thus, concentration due to residual salts in the soil or changes caused by additions of fertilizers, additions of irrigation water, utilization by the crop, or leaching can be correlated with crop response. Because salt concentration varies inversely with moisture, interpretation is dependent on a moisture level that can be used as a common denominator. Logically, this would be a field moisture level normally maintained for a growing crop. Irrigated sandy soils in Florida might contain 12.5% moisture at a good growing level. The saturation moisture content for these soils is about 25%; thus a saturation extract reading of 1,000 ppm could be calculated as 2,000 ppm at the 12.5% level and range on up to 4,000 ppm if the soil moisture decreased to 6.25%. The intensity and balance of most soil solutions, regardless of the different soils, are comparable with respect to a predictable growth response if the field moisture level is used as the common denominator. Water content as affected by both irrigation and drainage is not only relevant to interpretation but also to the associated production. Field moisture levels in different soil types or growing media vary widely, increasing with clay and/or organic matter content. By establishing a relationship between saturation and field moisture content, the measured conductivity of the saturation extract can be used to estimate the total salt concentration in the soil solution in the field.

Salt concentrations (ppm in the soil solution) as correlated with treatment and crop response are presented in Table 1. Nutrient and salt concentration, varied with scheduled fertilizer application which, in turn, were altered by periodic rains. With synthetic mulch (Geraldson, 1962, 1970) and a single application of surface banded fertilizer, concentrations decreased by gradient from the point of application and provided a mini-variable root environment compared to a maxi-variable environment without mulch. Crop response as measured by yield was increased by more

Table 1. Chronological variation in ionic concentration* in the soil solution from depth increments of the soil bed with and without plastic mulch and the associated yields (Experiment plots, spring 1962)

			ppm in soil solution at field moisture level					
			Fertilizer band			Plant row		
Mulch	Sample depth	Tomato yields	17/4	21/5	22/6	17/4	21/5	22/6
	cm	metric tons/ha						
–	0–5	48.5	950	12,300	300	900	5,800	1,240
	5–10		1,050	1,250	500	1,200	1,450	900
	10–20		1,210	950	920	1,190	930	1,050
+	0–5	67.0	27,000	19,800	12,500	6,000	7,400	13,500
	5–10		4,600	6,100	5,000	1,500	5,700	5,800
	10–20		1,500	2,300	3,100	950	2,600	2,300

* Equivalent quantities of fertilizer were banded 30 cm to either side of the plant row applied on the soil bed surface in one application with mulch, and weekly increments of the total without mulch. The bed was crowned toward the plant row and increments of the soil bed were sampled as indicated. Leaching rains preceded the 17 April and 22 June samplings. Seepage irrigation provided a subsurface source of moisture.

than 40% in this experiment by utilizing a combination of management components favoring minimal root environment stress. A soil or hydroponic solution that favors good root development might range from 1,000 to 3,000 ppm total salts. Portions of a soil or the total soil may contain an excess or deficiency as indicated by salt concentrations (Table 1). It should be obvious that these measures do fluctuate with additions of fertilizer, movement by water and utilization by crops and can be correlated with crop response. However, measures of nutrient intensity do not indicate levels of specific ions and a measure of nutrient balances within a range of intensity favorable to crop growth is necessary to predict crop response.

NUTRIENT BALANCE

Although salt concentrations are estimated from conductivities, individual ion contents are measured in the saturation extract and expressed as a proportionate part (%) of the total. Thus a normal percentage range for each cation and anion can be established for optimum crop response. Over a period of years such evaluations have resulted in a recorded series of ionic balances that range from deficient to adequate to excessive (Geraldson, 1962, 1963, 1967). A range of balance for most of the macro ions normally contained in the soil solution are presented in Figure 1. The levels of both macro- and micronutrients in the root environment portion of the soil solution would be comparable to that in the hydroponic solution. However, the latter media is homogenous whereas the soil solution is heterogenous with regard to both nutrient concentration and balance.

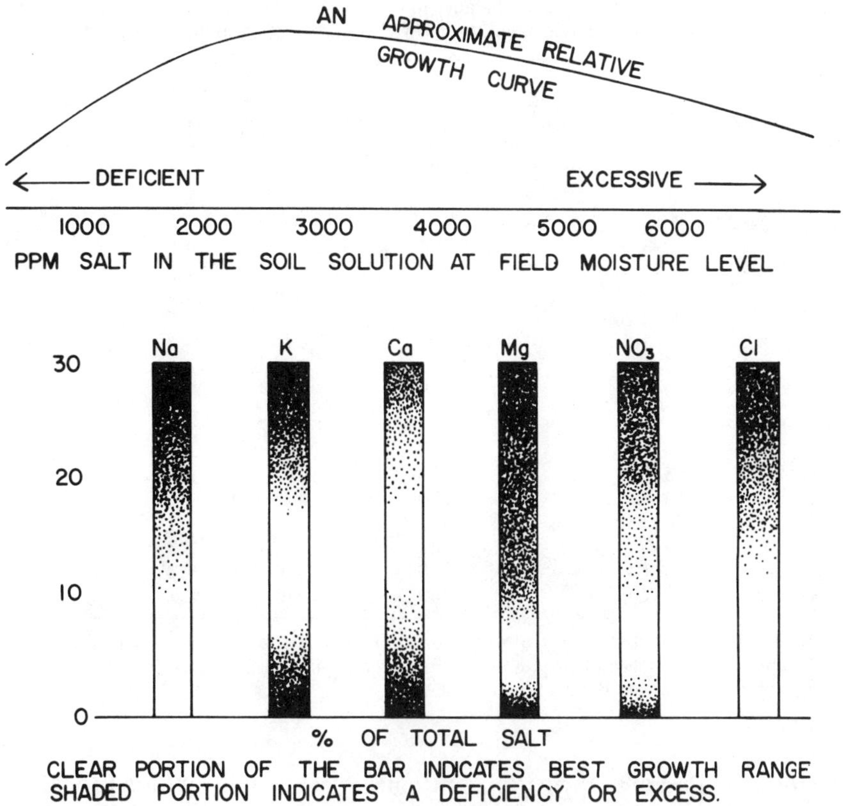

Figure 1. Range of nutrient concentration and balances associated with best production.

Thus, sampling and interpretation should be adjusted accordingly when considering correlations. (An indication of the variability that occurs is shown in Table 2 from samples collected from two areas in the same field.) As indicated in both Tables 1 and 2 utilizing mulches with one application of all the fertilizer banded on the surface results in a large degree of variability of the ionic concentration and balance in different portions of the soil bed. Relatively poor yields (28.8 metric tons/ha) from Area I (Table 2), can be correlated with high salts as well as proportionate excesses of sodium and chloride. The relatively good yields (123.4 metric tons/ha) from Area II can be correlated with the nutrient balance found in those portions of the soil bed where the concentration range (1,000-3,000 ppm) is most favorable.

The banded fertilizer contains primarily soluble nitrogen and potassium, thus these nutrients decrease proportionately with distance away from the point of application (Table 2). The proportions of all the nutrients are dependent on quantity and solubility of source materials. All of

Table 2. Ionic concentration and balance in the soil solution from depth increments of the soil bed and the associated tomato yields (Commercial field, spring 1967)

Field portion	Sample depth	Tomato yields	Salt	Percent					
				K	Ca	Mg	Na	NO_3-N	Cl
	cm	metric tons/ha	ppm	%					
I. Fert Band*		28.8							
	0-5		33,000	28.4	8.7	1.3	4.7	12.9	7.4
	5-10		11,200	17.3	5.0	1.1	17.9	4.5	23.2
	10-20		5,700	10.1	3.9	1.0	24.2	2.1	33.9
	Plant row								
	0-5		18,000	20.4	8.8	1.8	9.9	6.3	16.9
	5-10		11,500	13.7	6.1	1.5	18.8	6.5	25.0
	10-20		9,000	10.9	4.5	1.3	27.1	1.9	33.6
II. Fert Band		123.4							
	0-5		11,450	25.3	8.8	2.1	1.7	12.0	0.2
	5-10		2,500	12.0	10.3	6.9	9.5	11.9	1.1
	10-20		1,150	8.4	9.7	7.4	9.0	1.7	3.0
	Plant row								
	0-5		5,350	5.2	13.5	4.1	2.3	14.9	0.8
	5-10		4,250	4.8	11.9	5.2	7.3	6.0	0.5
	10-20		2,080	1.0	13.4	5.8	9.1	0.7	2.2

* Areas labeled I and II were from different portions of the same field with a major difference in growth and production. Fertilizer applied as indicated in Table 1 to a flat topped bed covered with mulch.

the ions equilibrate in the various increments of the soil solution providing ranges of concentrations and balances that can be manipulated toward an optimal for an associated optimal production.

Let us consider certain of the cations and anions that can be manipulated by fertilizer rate, source, ratio, and placement as well as movement with moisture.

Nitrogen

This nutrient can be supplied as a cation (NH_4^+) or an anion (NO_3^-). In the normal soil or hydroponic solution about 80 to 90% of the soluble nitrogen is present as nitrate, 10 to 20% ammonium. When a soil solution (root environment increments) contains larger proportions of ammonium, this cation can become competitive to the uptake of other cations and cause abnormalities (Geraldson, 1957b, 1967). There is also an anion antagonism as, for example, when chlorides become relatively high and nitrates low. Nitrates in the soil solution may vary from above 10% to zero, but if a gradient exists from a fertilizer band (Table 2) or if released from organics sufficient for the crop requirement, a reading of zero nitrate does not mean a nitrogen deficiency but would certainly indicate a potential deficiency. With reference to Field Area II, Table 2, it should be noted that with the gradient-mulch system (Geraldson, 1963, 1970) nitrates, as well as other nutrients in the root environment zone remain adequate to optimal, not deficient or excessive.

Calcium

In contrast to other macronutrients, calcium is not translocated from older plant tissue to younger, thus a calcium deficiency in the substrate is quickly reflected by abnormalities in the meristematic plant portions. Calcium uptake by the plant is affected by two basic factors (Geraldson, 1957b): one is the ratio or proportions of calcium in the soil solution. This can be altered by the addition of other soluble cations as well as calcium. The second factor is an excess of salt. As salt concentrations increase, the calcium becomes less available to the crop even though the ratio remains the same. Thus the available calcium by the normal testing procedure could be adequate to high and calcium could still be deficient because of an excess of a competitive cation or an excess of total salt, both of which could be diagnosed by the intensity and balance method as potentially affecting calcium uptake.

Applying sufficient limestone to Florida soils is basic to maintaining by equilibration an adequate ratio in the soil solution. Other calcium salts can be used to supplement that applied as lime. Excesses of competitive cations or total salts must be avoided.

Other Cations (Potassium, Magnesium, Sodium)

The range of sufficiency for these cations is included in Figure 1. In Florida, soluble potassium salts are applied as a major component of most fertilizers and a measure in the soil solution indicates the distribution in the soil bed from a point of application (Table 2). As with nitrogen, a low percentage in the root zone may indicate a potential deficiency, but only if the crop requirement exceeds that moving by gradient from a band in addition to that from the cation exchange or decaying organic matter in the soil.

In Florida, dolomite is a good source of magnesium as well as calcium. Magnesium is also applied in fertilizers as a soluble salt. Thus the quantity and distribution in the soil as measured in the soil solution would reflect that which may be applied in fertilizers or dolomite as well as that from the soil.

Sodium for the most part is not considered a plant nutrient but can accumulate to excess. Irrigation waters contribute varying amounts of sodium as do some fertilizers. Also, sodium chloride has been found to be a major residue salt from previous cropping. A residue salt can accumulate from decaying vegetation or fertilizer and/or irrigation water salts not utilized by preceding crops. With reference to Table 2 (Field Area I), the high sodium concentrations were attributed primarily to accumulation of residue salts.

Other Anions (Phosphates, Chlorides, Sulfates)

Only a small portion of the phosphates found in the soil is water soluble, thus that found in the soil solution usually measures less than 1%. Use of fertilizer phosphates for Florida sandy soils is generally more than adequate for phosphate nutrition.

In contrast, chlorides in the soil solution can range from 0 to more than 40% and, as with sodium, can come with irrigation water, some fertilizer salts, and as a residue in the soil (Table 2, Area I).

Sulfate is one of the three more soluble anions found in the soil solution, thus can be estimated as a reciprocal of the chloride and nitrate portion or it can be measured. The content of $(SO_4)=$ in the soil solution may vary, but normally would be less than 5 to 10%. Potassium, magnesium, sodium, and ammonium are the more soluble sources of sulfate; calcium sulfate is a less soluble source.

Minor Elements (Manganese, Zinc, Iron, Boron, Copper, Molybdenum)

The concentration of the minor elements in the soil solution ranges normally within 1 ppm and is comparable to that maintained in a hydroponic solution. Even though there is a narrow range of sufficiency, some excesses or deficiencies can be correlated with crop responses.

Correlations

The nutrient intensity and balance in the soil solution is a good measure of the equilibrium, which, in turn, reflects the relevance of the contributing components. Interpretation is dependent on correlations which are relevant to the heterogenous nature of the soil as well as the system of production that is applied to the soil. Theoretically, nutrient intensity and balance in the soil solution (root environment portion) at a field moisture level as well as that in the hydroponic solution are comparable, regardless of soil type or growing media.

From 1955 to 1963 tomato (*Lycopersicon esculentum*) yields in Florida increased about 50% from 800 marketable units (13.6 kg) per hectare to 1,200 (Rose, 1973). For the next decade yield increases were relatively static, but from 1973-75 yields increased markedly, an additional 60% to 1,920 marketable units/ha (Hawkins, 1975). The period of 1955-63 yield increase was associated with an effort to shift the soil root environment toward a more optimum nutrient intensity and balance particularly maintaining a higher calcium and nitrate balance as well as a more consistent and higher intensity. Because of periodic application of fertilizer, periodic leaching, as well as utilization by the crop, these levels were of a transitory nature. The 1973-75 increase was associated with an effort to maintain that optimum by means of the gradient-mulch system which primarily eliminated the transitory nature and minimized root environment stress. Production increases are at least in part a measure of the success of these efforts. A measure of the intensity and balance of nutrients in the soil solution has been used as a guide to establish, as well as maintain, an ionic root environment that can be correlated with optimal production. Approaching an optimal production by use of the gradient-mulch concept is a program based fertilizer management system rather than a soil test based one.

SUMMARY

The "nutrient intensity and balance" soil testing procedure measures the ionic equilibrium in the soil solution. The resultant has been used to support a management program designed to increase production of vegetable crops in Florida. The electrical conductivity of the saturation extract is used as an indicator of nutrient concentrations or intensity which can range from deficient to optimum to excessive for crop production. Specific cations or anions contained in the saturation extract are calculated as percent of the total salt concentration and used as an indicator of nutrient balance. Utilizing the nutrient intensity and balance measures as a guide, combinations of controllable factors (nutrient quantities, ratios, sources, placement, and movement) applied to soils can be used to shift the ionic root environment toward an optimum for plant growth and production. From 1955 to 1963 recommendations to establish a more favorable nutrient intensity and balance were associated with a 50% increase in tomato yields. However, established nutrient levels were of a transitory nature and for the next decade yield increases were relatively static. From 1973-75 a production management system designed to maintain the established root environment was associated with an additional 60% yield increase. These two periods of production increases were at least in part a measure of the efforts to establish and maintain a more optimal nutrient intensity and balance.

This testing procedure is being used by the state soil testing laboratory as well as a number of other laboratories in Florida primarily in the management of the more intensive systems of production. At present a Commission of the International Society of Horticultural Science is considering use of the saturation extract as a universal reference method for extraction of plant nutrients from the soil.

LITERATURE CITED

Brooke, D. L. 1975. Cost and returns of vegetable crops in Florida. Season 1973-74 with comparisons. Food and Resource Economics Department, Agric. Exp. Stn., Institute of Food and Agricultural Sciences, Univ. of Florida, Gainesville. Econ. Info. Rept. 22.

Geraldson, C. M. 1955. Soil solution soluble salts as an indicator of fertility level and nutrient balance. Soil Sci. Soc. Fla., Proc. 15:22-30.

Geraldson, C. M. 1957a. Soil soluble salts—determination of and association with plant growth. Fla. State Hortic. Soc. 70:121-126.

Geraldson, C. M. 1957b. Factors affecting calcium nutrition of celery, tomato and pepper. Soil Sci. Soc. Am. Proc. 21:621-625.

Geraldson, C. M. 1962. Growing tomatoes and cucumbers with high analysis fertilizer and plastic mulch. Fla. State Hortic. Soc. 75:253-260.

Geraldson, C. M. 1963. Quantity and balance of nutrients required for best yields and quality of tomatoes. Fla. State Hortic. Soc. 76:153-158.

Geraldson, C. M. 1967. Evaluation of the nutrient intensity and balance system of soil testing. Soil Crop Sci. Soc. Fla., Proc. 27:59-67.

Geraldson, C. M. 1970. Precision nutrients gradients—a component for optimal production. Commun. Soil Sci. Plant Anal. 1(6):317-331.

Hawkins, W. 1975. Florida Tomato Committee Tomato Review. June 1975. Florida Tomato Committee, Orlando, Florida.

Richards, L. A. (ed.) 1954. Diagnosis and improvement of saline and alkali soils. Agric. Handb. no. 60. USDA. U.S. Government Printing Office, Washington, D.C.

Rose, G. N. 1973. Tomato production in Florida—a Historic Data Series. Food and Resource Economics Department, Agric. Exp. Stn., Institute of Food and Agricultural Sciences, Univ. of Florida, Gainesville. Econ. Rep. 48.

Interpreting Inorganic Nitrogen Soil Tests: Sample Depth, Soil Water, Climate, and Crops[1]

C. M. Smith[2]

Nitrogen fertilizer recommendations for several crops are now based on soil tests for inorganic N particularly in many states of the Great Plains and the West. Nitrate (NO_3^-), and ammonium (NH_4^+) in some areas, measured on soil profile samples taken before crops are planted or in early stages of growth is useful in N fertilizer management for cereal grains, corn (*Zea mays* L.), sugar beets (*Beta vulgaris* L.), potatoes (*Solanum tuberosum*) and certain other irrigated and nonirrigated crops.

In most soils a high percentage of the total amount of N is in complex organic compounds. This organic N is not absorbed by plant roots and conversion to plant-available inorganic forms requires N mineralization. Rates of N mineralization depend on temperature, water, oxygen, and chemical and physical properties of the organic materials and the soils. Products of mineralization most readily utilized by plants are NO_3^- and NH_4^+.

In general NO_3^- moves in soils with water, can be leached beyond the rooting depth of plants, is subject to denitrification under certain conditions, and can be immobilized by microorganisms. Although NH_4^+ does not move readily in most soils, it converts to NO_3^- rapidly if the conditions are favorable for nitrification. Because of NO_3^- mobility traditional plow-depth soil sampling has produced inadequate data for N fertilizer recommendations in most cases.

Considering the dynamic nature of N compounds in soils and the mobility of NO_3^-, it is not surprising that soil test and recommendation programs for N have been difficult to develop. A thorough review of the history of biological and chemical soil testing procedures developed to determine N availability indexes has been made by Dahnke and Vasey

---

[1]Contribution from the Cooperative Extension Service, Montana State University, Bozeman.

[2]Professor and extension soil scientist. Presently, chairman and professor of soils, Department of Soils, North Dakota State University, Fargo.

(1973). Most of those procedures did not take into account initial or residual NO_3^-. These authors also reviewed the merits of soil tests for measuring inorganic N in soils at a given time.

The purpose of this paper is to present N soil testing and interpretation procedures for N fertilization presently being used in several states. Because the most current information on presently used practices is desired, many of the source references are extension circulars, semi-technical magazines, and personal communications from research or extension personnel.

Characteristics common to each of the N soil testing programs discussed include deep sampling, that is deeper than the traditional plow depth of approximately 15 cm, and major emphasis on measuring NO_3^-. In addition, some utilize stored soil water, anticipated growing season precipitation, NH_4^+, expected yield or yield goal, and other factors. Another refinement is a budget method of adding measured initial NO_3^- and NH_4^+ plus expected mineralization and subtracting crop requirements and anticipated losses or inefficiency of available N. The result is a rate of N fertilizer recommended for the crop.

A brief discussion will be given of soil sampling and sample handling for inorganic N testing as well as for soil water content. Also, some of the problems encountered in practice are discussed. Correlation of soil test data with plant responses to N fertilizers and subsequently calibration of lab tests with N-rate experiments in the field are assumed as adequate. Such procedures are discussed earlier in chapters by Melsted and Nelson.

SOIL SAMPLE DEPTH

Depth of soil sampling needed to evaluate inorganic N can vary with soil texture, soil depth, climate, irrigation,and crops to be grown. Among several states, recommended soil sample depths range from 60 to 180 cm.

Nonirrigated Crops

If the goal is to evaluate NO_3^- and available water supply to the depth of significant water removal, for example by a dryland winter wheat (*Triticum aestivum* L.) crop, it may be necessary to take samples to a depth of 180 cm according to research in Montana and Washington (Brown, 1971; Halvorson, 1971; Leggett, 1959). However, there is evidence in areas of the Great Plains from Oklahoma to Saskatchewan that a 60 cm soil sampling depth usually produces data that evaluate inorganic N status with acceptable precision for their soil and climatic conditions for cereal grains (North Dakota Cooperative Extension Service, 1975; Read & Warder, 1973; Saskatchewan Advisory Council on Soils, 1975; Tucker, 1975; Wagner et al., 1974; Whitney, 1972, 1974).

Nebraska (Sander et al., 1973) indicates a preference for 180 cm sample depth for winter wheat. However, sampling to that depth is generally not done by farmers due to unavailability of hydraulic equipment or commercial services. In practice, because soil samples received in the laboratory have been taken to variable depths, data are adjusted for purposes of recommendations to a common 180 cm, or depth of soil. This is based on an average of NO_3^- distribution data in soil profiles of numerous field experiments conducted in the area. They believe resulting errors are less than if no adjustments are made. Some of the other states that advocate deeper depths for dryland crops are Idaho (J. P. Jones, personal communication. University of Idaho) and Utah (R. E. Lamborne, personal communication. Utah State University), Washington (Engle et al., 1975), and Montana (Christensen & Smith, 1973; Smith, 1970). Recommended sampling depths vary from 90 to 180 cm, or to an obstructing layer for roots. All states agree that NO_3^- analysis of only a plow depth sample has little value for making N recommendations.

Data about distribution of NO_3^- in soil profiles exist for many different areas, soils, and management systems. In general, for a given climatic region, limited fertilization and relatively uniform soil and crop management produce similar amounts of NO_3^- in soils below 60 cm deep. However, exceptions do exist and as use of N fertilizers increases along with some NO_3^- movement into subsoils a greater need develops for deeper soil sampling. Variability in NO_3^- distribution is further emphasized especially where precipitation is relatively nonuniform within areas and among years.

Daigger, et al. (1972) in an experiment in western Nebraska incorporated ammonium nitrate into a fallowed soil. At the N rate of 180 kg/ha, for example, and after 22 cm of rain in 127 days, NO_3^- concentration increased by 10 ppm N between zero and 15 cm deep, and by 9 ppm N between 60 and 120 cm deep. The least change of about 4 ppm N was between 15 and 60 cm deep. It should be noted that a different amount of precipitation probably would have changed the profile distribution of NO_3^-. Ammonium concentrations in soil samples changed significantly only in the zero to 15 cm depth, but not until rates of N fertilizer application reached 180 kg/ha. In this experiment zones of maximum and minimum concentrations of NO_3^- were independent of N fertilizer rate.

In a field in eastern Montana that had never been fertilized with more than 4 kg/ha N per 2 years, the author measured NO_3^- concentrations greater than 100 ppm N between 120 and 150 cm deep, well within reach of wheat roots. However, in the upper 60 cm of depth this same soil averaged less than 4 ppm N as NO_3^-. Distribution patterns with significant NO_3 concentration bulges below 60 cm of depth are not uncommon in some areas. Knowledge of the expected profile distribution of NO_3^- is essential before recommending the depth to sample. Also, the location of plant available N within the profile can have a major affect on the sufficiency or deficiency of N at certain stages of physiological development of plants.

Irrigated Crops

Soil sampling depths recommended for irrigated sugar beets, corn, or other grain crops are usually 120 to 180 cm (Carter et al., 1974; Christensen et al., 1974; Daigger et al., 1973; Dow et al., 1969; Gardner, 1971; Giles et al., 1975; Ludwick et al., 1975; Wiese & Penas, 1974), however some use 60 cm or less as of this date (Ludwick et al., 1975; North Dakota Cooperative Extension Service, 1975). Because of the shallow rooting of potatoes, recommended soil sampling depths are 60 to 90 cm (Dow et al., 1969; Gardner, 1971; Painter et al., 1975). Sampling soils to greater than rooting depth for such crops as potatoes may not help to assess the amount of NO_3-N available for the crop. But these deeper samples can provide valuable data about NO_3^- movement in the soil and about water and N fertilizer management. For example, if excessive NO_3-N from fertilizers has moved beyond the root zone, reduced amounts of irrigation water per application and more frequent irrigations may be necessary. Also, changing fertilizer N management to split applications could be practiced to reduce leaching.

Profile distribution studies after fertilizer N application and irrigation treatments indicated NO_3^- concentrations often increased at depths greater than 120 cm (Anderson et al., 1972; Ludwick et al., 1974). Experiments in Nebraska (Anderson et al., 1972) using labeled N (^{15}N) produced data showing sugar beets absorbed N from the deepest placement of 135 cm. Also, beets grown in a soil testing high in NO_3^- absorbed less deep NO_3^- than from a soil testing low. This has important implications in N management for sugar beets. For example, if sampling were only to 60 cm deep and a low soil test resulted in application of high rates of N fertilizer, lowered sugar production could result if there were significant amounts of deeper residual NO_3^- within reach of roots.

Deep sampling is also useful for identifying fields with high levels of NO_3^- so crops such as malting barley (*Hordeum vulgaris* L.) and soft wheats requiring low protein can be grown elsewhere. Also, optimum N rates are easier to determine for irrigated malting barley according to research by the author in Montana.

SOIL WATER AND PRECIPITATION

The success of dryland or nonirrigated crop production is dependent on precipitation during the crop growing season and plant available water stored in the soil. It is possible to determine soil water in late winter or early spring. But often it is not easy for farmers to quantitatively measure it. Expected precipitation is also difficult to estimate. However, long time records have been analyzed and guidelines prepared to assist with N fertilizer recommendations, as has been done in North Dakota (Wagner &

Vasey, 1971). Farmers are encouraged to keep precipitation records (Smith, 1970) and compare with official weather stations. They soon realize how their farming areas differ.

Tucker (1975) stated the probability of achieving a certain dryland grain yield in the High Plains of Oklahoma depends upon the total available water for the crop. In one study plant available soil water at seeding plus precipitation during the growing season was highly correlated with wheat yields for the years 1957 through 1963. The high risk of failure when dryland crops must depend upon growing season precipitation is greatly reduced as stored soil water at time of seeding is increased. Knowledge of stored soil water, then, is a major factor in yield potential and hence affects the decision about how much N fertilizer to apply.

Brown (1971) in Montana has shown the interrelationship of soil water, growing season precipitation, and N levels for producing high yields of grain. He related wheat plant development to a continuous water budget during its growing season. Adequately fertilized winter wheat grown on a medium textured soil extracted more water from intermediate soil depths than N deficient treatments. Sufficient N also caused root extension and water extraction to greater than 180 cm. This emphasizes the need to know soil water status for the potential crop rooting zones.

In practice a special effort is needed to obtain soil water data with adequate equipment being required for deep sampling. For the data to be meaningful the best time to assess soil water status is near fertilization, and preferably as near the growing season as is practical. Gravimetric determinations of water are usually made on the same samples as NO_3^-; however, if sampling is done in the fall an estimation of water storage in the soil from winter precipitation will be needed. Custom sampling as a service to farmers is discussed later. Such a service is essential before sampling for either NO_3^- or water can be successful. Even with custom services high precision of measuring plant available water in soils is difficult on a field scale. However, when considering the level of precision associated with predicting growing season precipitation, assessing other variables that affect yield, and the difficulty of obtaining accurate soil bulk densities, adequate estimates of soil water may be practical.

An alternative method for estimating soil water is to use a probe (Brown, 1959) to measure depth of moist soil. Then an estimate of plant available stored water can be made by determining texture and by applying appropriate factors. This gives useful "approximate" values (Brown, 1959; Cassel & Sweeney, 1974; Christensen & Smith, 1973; Smith, 1970; Wagner & Vasey, 1971). The farmer or farm adviser can combine this information with expected precipitation for better guidelines to adjust N rates than to guess the plant available water situation. This estimate of stored soil water is also useful if expected yield is a basic factor in the recommendation system.

Another procedure for estimating the effect of probable available water for a dryland crop is to establish a yield goal with adequate fertilization (Wagner et al., 1974). If a yield goal for dryland grain, for example, is realistically set it should serve as an integrator of expected precipitation and stored soil water that the farmer believes will occur for his farm. The best source of yield goals is fertilizer rate experiments conducted under optimum management and for a number of years. However, such data are not available for all soil and climatic conditions.

Farmer yield records can be a guide for establishing goals but should be used carefully. They, however, reflect the farmer's management ability. Expected yields should not be disproportionate to yield potentials based on experimental data. Used wisely, this system can be a major improvement over using NO_3^- as a single variable. Many states use yield goals in combination with inorganic N by soil test as a basis for adjusting N fertilizer rates. This in effect becomes a budget approach to recommendations.

Seeding dryland crops on stubble land rather than on summer fallow is increasing in much of the northern Great Plains and the Pacific Northwest. Because of greater mechanization and generally improved soil management, farmers are becoming more efficient in conservation of water in the soil. Also, techniques of snow trapping in standing stubble can increase water storage over winter and improve the chances of producing a satisfactory yield without summer fallow. With this changed cropping sequence, use of soil water measurements and growing season precipitation probabilities increases in importance. Most states have or can obtain precipitation probability information for the crop producing areas. This can be developed for a 4-month growth period or for whatever frequency is desired. More crop yield and N response data as related to precipitation and precipitation probabilities are being acquired by researchers in many areas to improve understanding and usefulness of this approach.

USING THE SYSTEM

There is no single best approach to obtaining or using data about inorganic N for N fertilizer recommendations. Nitrogen availability to crops and its effectiveness in producing yield increases is linked so closely to plant available soil water, precipitation, temperature, and to other factors that one should expect system modifications in different regions. If the amount and distribution of NO_3^- would be reasonably constant in a soil management system, soil testing would be needed only frequently enough to check the status. But that is usually not the situation.

Making use of this system of deep testing is discussed in three parts: taking samples correctly and handling them so minimum changes occur, uniformity in analysis and reporting of results, and interpretation of test results and procedures used for making recommendations. Some aspects of these vary markedly among states.

Taking and Handling Soil Samples

The one greatest problem associated with farmer participation in deep soil sampling is ability to take the samples and subsequent handling of them, according to those in charge of either the state soil testing laboratories or of making fertilizer recommendations in the Great Plains and Northwest states. They indicate that farmers are willing to pay for custom services when they are available. Progressive fertilizer dealers in some states are providing a sampling service to customers. This is sometimes free if the fertilizer is purchased from that dealer, or a fee may be charged. Custom services are also available from private soil testing and farm management firms in states where deep testing is being used to a significant extent. In North Dakota, for example, custom sampling is a profitable sideline business for a rural mail carrier and for many others. Dahnke believes one reason for an extremely rapid increase in soil testing for NO_3^- in North Dakota is availability of custom sampling (W. C. Dahnke, personal communication. North Dakota State University.).

Deep samples are not hard to collect with vehicle-mounted hydraulic soil samplers. Such equipment is available commercially and is relatively inexpensive for depth capability of about 60 cm. However sampling to 120 or 180 cm requires heavier and more costly equipment. Special designs of slotted and larger diameter tubes make sampling easier if soils are high in clay content and sticky when wet, or when soils contain gravels that can easily plug small diameter probes. Open faced hand probes are widely used in medium to moderately coarse soils, but extreme care is required to avoid contamination. Augers are also available but should be used only in extremely difficult sampling conditions as contamination is a serious problem. Although sampling by hand is practical, the number of soil samples taken is likely to be inadequate.

Consider the area of a field to be represented by a composited set of profile samples as a "sampling unit." Each sampling unit should represent areas of similar soil types with past crop and fertilizer management uniform for a minimum of 3 to 5 crop years. For a composite sample (Northwest Soil and Plant Test Work Group, 1974; Smith, 1972) representing the surface to 15 to 30 cm in the sampling unit, 15 to 25 cores should be mixed together. Fewer cores are needed from deeper depths. Unless the subsoils are extremely variable, the number of cores for the second sample depth or to 60 cm can usually be reduced by one-third or to 10 or 15 cores. Generally less variability in inorganic N occurs at depths greater than 60 cm so a composite of 5 to 7 cores per sample is sufficient. Most states advocate separate analysis for each 30 cm to obtain NO_3^- distribution. States advocating 60 cm sampling usually suggest plow depth or 15 cm as one sample and from that depth to 60 cm for the second. Each sample is analyzed separately because in addition to NO_3^- the plow depth sample may be analyzed for NH_4^+, pH, P, K, other elements, organic matter (OM), and soluble salts.

Soluble salts, including nitrates, are not uniformly distributed in soil after rill or furrow irrigation and special procedures are recommended to avoid nonrepresentative samples (Dow et al., 1969; Northwest Soil and Plant Test Work Group, 1974; Smith, 1972). If the row direction is known, a set of three cores per depth should be taken in an orientation perpendicular to the row or ridge direction and spaced at one-fourth the distance between ridges. Repeat this procedure for each 0.8 to 1.2 ha (2 or 3 acres) and composite by depths the sets of 3 to make up a sampling unit. Each set of three cores can be composited and analyzed separately if desired. Or, when sampling is to determine NO_3^- distribution in furrows or ridges, compositing can be according to position as well as depth assuming exact location of furrows is known.

After taking samples for inorganic N analysis biological activity in the soil must be stopped or retarded. If this is not done analysis for NO_3^- and NH_4^+ may yield incorrect data (Northwest Soil and Plant Test Work Group, 1974; Smith, 1972). The author subdivided soil samples and measured increases in NO_3^- of 10 ppm N or more in field moist soil samples that were transported 2 days as compared with counterpart samples that were frozen after being taken. Therefore, to avoid significant changes, samples should be spread in a thin layer and drying initiated within a few hours after sampling, or samples could be frozen or refrigerated until drying. When drying the soil samples forced air and temperatures at less than 50°C should be used. Plastic lined bags set into an oven retard drying rate so much that microbial activity can produce significant changes in forms of N. High clay content of soil, large samples, and not using forced air cause many problems associated with drying.

Problems of contamination in the field are essentially the same as for samples taken for any soil tests. However, if analytical tests include NH_4^+, measures must be taken to avoid contamination of samples by ammonia (NH_3). Samples held in storage, even though dry, can adsorb enough NH_3 to produce large errors.

Analysis and Reporting

States and provinces differ in their use of NO_3^-, or both NH_4^+ and NO_3^-, as a measure of plant available N in soil samples. Reports from Kansas (Whitney, 1974), Idaho (Painter et al., 1975), and Saskatchewan (Saskatchewan Advisory Council on Soils, 1975) indicate NH_4^+ is of sufficient quantity in soils, especially in the spring, to be considered significant for recommendations, although the University of Saskatchewan laboratory does not routinely test for NH_4^+ (J. Hamm, personal communication. University of Saskatchewan). Generally when both ionic forms of N are measured they are considered as equal in availability to plants and are summed when making recommendations. Most states that only test for NO_3^- have had experiments where NH_4^+ was of significance. However, NH_4^+ concentrations were reasonably constant and are not used as a variable when making recommendations.

Reporting of data is not consistent among labs, which can cause confusion for those who read or use laboratory reports from different areas. The concentrations of NO_3^- and NH_4^+ measured in extractions from soil samples are apparently expressed as N in all cases. Differences of expression caused by calculation procedures and factors are imposed subsequently. Some soil test reports show N as parts per million. Others convert concentration (ppm) to quantity (pounds per acre) before reporting, or both may be given.

Some labs add all concentration values for the profile to obtain a "total" ppm. The reader is then instructed to multiply four times the "total" ppm of N for all depths to obtain total pounds per acre. It is assumed all soil samples are by increments of 30 cm (12 inches). This is a questionable practice because different sample depths are analyzed and often are not weighted, which produces errors. Also, the "total" value obtained by adding is not correct for the concentration of the whole profile. An average would be correct for the profile if concentrations were weighted by depths. But, by using this procedure the advantages of knowing profile distribution are lost. Using a sum of concentrations reported as ppm of N is misleading even if interpretation tables are labeled appropriately. The tables in fertilizer guides often are for a fixed profile depth, but the actual sample depths may add to more or less than that.

Some states use 4 million lb/acre per foot (4.5 million kg/ha per 30 cm) of soil depth and convert ppm of N to lb/acre (lb/acre $\times$ 1.12 = kg/ha). This means they assume the soil bulk density is constant for all depths and for all soils. The assumption is that errors introduced in this way are not as serious as others in the system. This assumption is probably correct until precision is improved for other variables.

The several reporting procedures described above are intended for easier use of soil test data. Attempts to oversimplify have in some instances contributed to computation steps that are not only confusing but sometimes erroneous. If NO_3^- and NH_4^+ concentrations were reported as ppm of N for the specific sample depth, a simple equation could then convert the data to quantity (kg/ha or lb/acre) if desired. Or, both the ppm of N and the calculated kg/ha or lb/acre can be printed out for each depth. Providing correct information for each depth gives the purchaser of the soil testing service the best data.

Interpretations and Recommendations

Making N fertilizer recommendations from soil test data for inorganic N, ranges from using equations that incorporate several variables to availability index ratings of very low, low, medium, and high with associated N fertilizer rates for different crops. Most widely used procedures fall in between, but variations in simplicity can even occur among crops within a state. Examples of procedures used by some states for certain crops are given. These are selected to show different techniques being used for N recommendations, and some of the examples may be out of date soon.

Many researchers have examined relationships between soil test data for inorganic N measured before the crop was grown and yield responses to N fertilizer (Dahnke & Vasey, 1973). A general summarization is that high correlations are not uncommon for single experiments, but they gradually drop as additional experiments are added even for the same soil series and year. The correlations become poorer as experimental data from more soils and climatic conditions are added.

In the mid 50's Leggett (1959) in Washington showed a good relationship between dryland wheat yields and available water and nitrogen. This research plus additional more recent data has evolved into the present fertilizer guide for dryland winter wheat in eastern Washington (Engle et al., 1975). It is a budget type of approach.

BUDGET METHOD

This system utilizes a computation of the amount of N required to produce an estimated potential yield based on response data from experiments. Then an inventory of the plant available N from all sources other than fertilizers is made. The plant available N is subtracted from the total N required for the potential yield, with the result being the N needed as fertilizer. There are several modifications of this system in use.

Washington Dryland Winter Wheat recommendations (Engle et al., 1975) are developed in four steps: (i) estimate potential yield from plant available water, (ii) calculate N needed to obtain that yield, (iii) inventory the soil N, and (iv) calculate needed N fertilizer rate. More specifically this is done as follows:

i. Potential yield = (yield/unit of water) × (expected precipitation plus available soil water minus water to grow plant)

Yield(hl) = 0.97 hl/cm (ppt, cm + soil water, cm − 10 cm)

Or, Yield(bu) = 7 bu/inch (ppt, in + soil water, in − 4 inches)

Note: For soft white winter wheat use 0.97 hl/cm (7 bu/inch) and 0.83 hl/cm (6 bu/inch) for hard red winter wheat.

Precipitation (ppt) is that expected for wheat growing season.

Soil water is plant available water to 180 cm or to a root restricting layer.

10 cm of water is used to grow the plant.

ii. N needed (kg/ha) = ___(i)___ hl/ha × 3.5 kg/hl

Or, N (lb/A) = ___(i)___ bu/acre × 2.7 lb/bu

Note: Use 3.5 kg/hl of N (2.7 lb/bu) for soft white winter wheat and 3.9 kg/hl of N (3.0 lb/bu) for hard red winter wheat

iii. Soil N inventory = soil test NO_3^- as kg/ha of N (or lb/acre) to 180 cm *plus* expected N mineralized during the growing season from OM, previous crop if legumes, and other sources *minus* loss of available N by immobilization, volatilization, denitrification, and leaching.

iv. N fertilizer rate = (ii) − (iii)

North Dakota Recommendations for N fertilization of dryland small grains (Wagner et al., 1974) are determined by a modification of the budget approach. A yield goal is requested from the producer who is asked to consider the highest yield he has had for the particular field. He can make adjustments in yield goal depending on data from N-rate experiments, stored soil water, expected precipitation, and other management factors. A warning is given that unrealistic yield goals will result in incorrect N recommendations.

The N required to produce a crop depends on yield, at least within certain limits. The guide includes a chart of yield goals and N requirements for wheat, durum, rye (*Secale cereale*), feed barley, and oats (*Avena sativa* L.) developed from numerous field experiments. The difference between total N required and N available in the soil to 60 cm is the N fertilizer recommendation.

They also provide the producer or the adviser with a circular (Wagner & Vasey, 1971) that relates soil water and precipitation to expected yield responses to fertilizers. Instructions are given on how to estimate plant available soil water by depth of wet soil and soil texture. A map is also provided that shows percentage probabilities of receiving greater than 15 cm (6 inches) of precipitation for the period 3 May through 1 August, a 13-week period during which precipitation has a major influence on grain yield. North Dakota's use of yield goals gives the responsibility for determining a realistic yield to the farmer. This is where the responsibility should be and it works well unless his estimates are too high due to incorrect assessment of plant available water or of other factors affecting potential yield.

Montana Irrigated Sugar Beet Recommendations for N fertilizer are given in a fertilizer guide (Christensen et al., 1974) and are based on budget inputs of: (i) total N requirements, (ii) soil test NO_3^-, (iii) estimated mineralable N, (iv) N from manure, and (v) N from previous crop. These are combined to obtain the N rate to apply.

These budget inputs are determined as follows:

i. Data from several N-rate experiments indicate an N requirement for sugar beets of about 4.5 kg/metric ton (9 lb/ton). The yield goal is the 5-year average for the field. This average should reflect the producer's management for the soils in the field and the growing season conditions. The sugar beet producer is reminded that excessive N whether residual or from fertilizer can reduce sugar production even though total dry matter yield may be increased.

ii. Soil test data are for NO_3^- to a depth of 120 cm. Warning is given about obtaining soil samples that are representative of conditions. Also an equation is given for quantitative calculation of N from concentrations (ppm) given in the soil test report and any depth of sample.

iii. N mineralization is estimated as about 34 kg/ha (30 lb/acre) per 1% OM, during the growing season. This is an average of data from several experiments and could be different in other climatic areas.

iv. Manure contributions to N supply for sugar beet yield and sugar production are based on long time experiments at Sidney, Montana (Halvorson & Hartman, 1975). The available N per ton of manure is credited as 4 kg/metric ton (8 lb/ton) of N as available the first year and half that amount the second year.

v. Previous crop adjustment reduces the needed N fertilizer rate if that crop was legumes and increases the N rate when straw or corn stover is to be plowed down. The adjustment for rate of N fertilizer per ton of residue is 10 kg/metric ton (20 lb/ton).

DIRECT RESPONSE METHOD

Some states have related the yield response to N fertilizer to soil test N, soil water, and precipitation. Or, they have developed different tables of N rates for sections of their state. By this technique they reflect the climatic conditions, by areas, that existed when the experiments were conducted.

Montana Dryland Winter Wheat N fertilizer recommendations, as one example, are made from a fertilizer guide sheet (Christensen & Smith, 1973) that considers soil test NO_3^- and expected plant available water. The recommendations for soil samples are to 120 cm deep for NO_3^- analysis. Plant available soil water is determined either by gravimetric soil analysis or estimated from depth of moist soil and soil texture.

The grain producer is encouraged to keep precipitation records and relate them to those for the nearest weather station. A two-way table is provided that gives the N fertilizer rates recommended for each of several ranges of NO_3^- in the soil profile, at different amounts of plant available water of 20 to 36 cm (8 to 14 inches) and over. The latter are the sums of soil water and expected growing season precipitation. The highest N rate recommended is 134 kg/ha (120 lb/acre) for a NO_3^- soil test level of less than 22 kg/ha of N (20 lb/acre) to 120 cm soil depth and 36 cm (14 inches) or more of soil water plus expected growing season precipitation.

SUMMARY

Nitrogen fertilizer recommendations are being made for several crops based on soil tests for inorganic N in soil profile samples, plant available water for nonirrigated crops, and other management factors. Major use of N soil tests on samples taken to 60, 120, or 180 cm depth is in the Great Plains and the West.

Procedures may have been developed earlier for dryland crops in most of these areas because of the strong interdependence of plant growth response to N and to plant available water. Also, in these areas conditions are favorable for a buildup of residual NO_3^- because of summer fallow practices and the infrequent flushing of water through the soil profile. Sugar beets and oil seed crops, where quality as well as yields of sugar and oil are closely related to available N supply, have received much attention in development of inorganic N testing. Not only is it important to determine quantity of NO_3^- in the soil but its distribution is needed as well. Nitrate-N deep in the soil can be absorbed by sugar beet roots late in the season, for example, and can reduce total sugar production. Soil testing for NO_3^- has proved useful for N and water management on irrigated potatoes and other crops.

There is much variation among states and Canadian provinces as to depth for sampling, methods of reporting test data, variables used for making recommendations, and inclusion of NH_4^+. Regional efforts are underway in the Northwest and in the Great Plains to resolve some of these differences. Generally the approaches vary because of soil and climatic conditions that existed when experiments were conducted in each area. It is more difficult to integrate approaches when climatic and management factors are involved in recommendations.

There are major efforts underway to improve field calibrations for inorganic N soil testing, and to examine carefully whether a system can be developed for more crops in areas of present use and for certain soil and crop conditions in more humid areas. Collection and refinement of additional data, and imagination in its use, are needed to combine measurable and predictable factors into the most convenient and useful N fertilizer management techniques. Opportunities exist to improve N recommendations for many crops through timely soil testing for NO_3^- and NH_4^+ in soil profile samples. Goals are to improve crop product yield and quality in addition to optimizing N fertilizer use efficiency for energy conservation and environmental quality. In all cases it is apparent that more information is being obtained and used by those making N fertilizer management decisions.

LITERATURE CITED

Anderson, F. N., G. A. Peterson, and R. A. Olson. 1972. Sugar beet quality, soil testing can help. Nebraska Coop. Ext. Ser. QR 7-72.

Brown, P. L. 1959. Soil moisture holds the key! Crops Soils. 11(9):16-17.

Brown, P. L. 1971. Water use and soil water depletion by dryland winter wheat as affected by nitrogen fertilization. Agron. J. 63:43-46.

Carter, J. N., M. E. Jensen, and S. M. Bosma. 1974. Determining N fertilizer needs for sugar beets from residual soil nitrate and mineralizable N. Agron. J. 66:319-323.

Cassel, D. K., and M. D. Sweeney. 1974. In situ soil water holding capacities of selected North Dakota soils. North Dakota Agric. Exp. Stn. Bull. 495.

Christensen, N. W., D. E. Baldridge, G. P. Hartman, and A. D. Halvorson. 1974. Sugar beets — Irrigated. Montana Coop. Ext. Ser. Fertilizer Guide. AG55.610:60.

Christensen, N. W., and C. M. Smith. 1973. Winter Wheat — Nonirrigated. Montana Coop. Ext. Ser. Fertilizer Guide. AG55.610:21.

Dahnke, W. C., and E. H. Vasey. 1973. Testing soils for nitrogen. p. 97-114. *In* L. M. Walsh and J. D. Beaton (ed.) Soil Testing and Plant Analysis. Soil Sci. Soc. of Am., Madison, Wis.

Daigger, L. A., F. N. Anderson, and D. Knudsen. 1973. Fertilizing sugar beets. Nebraska Coop. Ext. Ser. NebGuide. G 73-8.

Daigger, L. A., D. H. Sander, and G. A. Peterson. 1972. Applied nitrogen—where is it? Univ. of Nebraska Coop. Ext. Ser. QR 5-72.

Dow, A. I., D. W. James, and C. E. Nelson. 1969. Interpretation of soil test N for irrigating crops in central Washington. Washington Coop. Ext. Ser. Cir. E. M. 3076.

Engle, C. F., F. E. Koehler, K. J. Morrison, and A. R. Halvorson. 1975. Dryland wheat nitrogen needs for eastern Washington. Washington Coop. Ext. Ser. Fertilizer Guide. FG 34.

Gardner, E. H. 1971. OSU Soil test for nitrate nitrogen. Oregon Coop. Ext. Ser. Cir. 770.

Giles, J. F., J. O. Reuss, and A. E. Ludwick. 1975. Prediction of nitrogen status of sugar beets by soil analysis. Agron. J. 67:454-459.

Halvorson, A. D., and G. P. Hartman. 1975. Manure good source of N for beets. Mont. Farmer Stockman. 62(9):21-23.

Halvorson, A. R. 1971. Soil moisture and nitrate tests—When are they useful? Washington Coop. Ext. Ser. Cir. E. M. 3523.

Leggett, G. E. 1959. Relationship between wheat yield, available moisture and available nitrogen in eastern Washington dryland areas. Washington Agric. Exp. Stn. Bull. 609.

Ludwick, A. E., J. O. Reuss, and E. J. Langin. 1974. Distribution of soil nitrate N following 4 years continuous corn affected by N rates and irrigation treatments. Colorado Agric. Exp. Stn. Res. Rpt. PR 74-35.

Ludwick, A. E., P. N. Soltanpour, and J. O. Reuss. 1975. Guide to fertilizer recommendations in Colorado. Colorado Coop. Ext. Ser.

North Dakota Cooperative Extension Service. 1975. 1976 Crop Production Guide: Diagnostic methods and fertilizer recommendations. North Dakota Ext. Ser. Cir. A-592, 44-61.

Northwest Soil and Plant Test Work Group: Subcommittee on Soil Sampling. 1974. Principles of soil sampling for Northwest agriculture. p. 115-126. *In* A. I. Dow (ed.). Proceedings Twenty-fifty Annual Pacific Northwest Fertilizer Conference. Richland, Washington. 16-18 July. Northwest Plant Food Assn., Portland, Oregon.

Painter, C. G., J. P. Jones, R. E. McDole, R. D. Johnson, and R. E. Ohms. 1975. Potatoes: Idaho fertilizer guide. Idaho Coop. Ext. Ser. and Agric. Exp. Stn. Current Information Series 261.

Read, D. W. L., and F. G. Warder. 1973. Influence of soil and climatic factors on fertilizer response of wheat grown on stubble land in southwestern Saskatchewan. Agron. J. 66:245-248.

Sander, D. H., L. A. Daigger, and G. A. Peterson. 1973. How much fertilizer on wheat? Nebraska Coop. Ext. Ser. NebGuide G 73-37.

Saskatchewan Advisory Council on Soils. 1975. Saskatchewan fertilizer and cropping practices 1975. Extension Division, Univ. of Saskatchewan, Saskatoon, Canada.

Smith, C. M. 1970. Fertilize winter wheat. Mont. Farmer Stockman. 57(23):7-9.

Smith, C. M. 1972. Diagnostic Methods. p. 115-141. *In* W. C. White and D. N. Collins (ed.) The Fertilizer Handbook. The Fertilizer Institute, Washington, D. C.

Tucker, B. B. 1975. Balancing fertilizer application to the limiting nutrient. Part 3: The High Plains. Fert. Solutions. 19(2):79-90.

Wagner, D. F., and E. H. Vasey. 1971. Refining fertilizer recommendations by state areas, stored soil moisture conditions and rainfall probabilities. North Dakota Coop. Ext. Ser. S & F No. 1.

Wagner, D. F., W. C. Dahnke, A. Bauer, and E. H. Vasey. 1974. Fertilizing small grains. North Dakota Coop. Ext. Ser. Cir. S-F 2.

Whitney, D. A. 1972. What about soil testing for available nitrogen? Fert. Solutions. 16(3):58-59.

Whitney, D. A. 1974. Soil test interpretations and fertilizer recommendations. Kansas Coop. Ext. Ser. Cir. C-509.

Wiese, R. A., and E. J. Penas. 1974. Fertilizer recommendations for corn. Nebraska Coop. Ext. Ser. NebGuide G 74-174.

Micronutrient Soil Test Correlations and Interpretations

J. J. Mortvedt[1]

Micronutrient deficiencies in agronomic and horticultural crops are widespread but are not always easy to diagnose. Increased crop yields resulting from improved cultural practices may increase the incidence of these deficiencies. More attention is thus being placed on efforts to improve methods for diagnosis and correction of micronutrient deficiencies.

Both soil tests and plant analyses have been used to diagnose micronutrient deficiencies in crops. Micronutrient concentrations in plant tissue commonly reflect amounts of available micronutrients in the soils. Interpretation of these results requires knowledge concerning interactions with other nutrients; physiological age, size, and plant part; crop species and variety; and other related parameters. Most soil tests for micronutrients have not been widely adapted and analytical results may be difficult to interpret. Therefore, some argue that plant analysis is better than soil testing as a diagnostic tool for micronutrients, although soil test results are also used in interpreting plant analysis results.

Soil tests for micronutrients have been reviewed and discussed at several symposia in recent years. Viets (1967) and Reisenauer (1967) discussed soil tests for micronutrient cations and anions, respectively, in 1965. Cox and Kamprath (1972) reviewed the literature on micronutrient soil tests at the micronutrient symposium at TVA in 1971. The most recent discussions of micronutrient soil tests are by Viets and Lindsay (1973) and Reisenauer, Walsh, and Hoeft (1973).

The above references include concepts and philosophies of soil tests, as well as a review of literature of the main micronutrient soil tests in use today. The purpose of this paper is to analyze problems and concepts involved in calibration and interpretation of soil tests. Some micronutrient soil tests are also described as illustrations and to suggest needed research;

[1]Soil chemist, Soils and Fertilizer Research Branch, Tennessee Valley Authority, National Fertilizer Development Center, Muscle Shoals, Alabama.

the references cited earlier give a more complete discussion of each soil test.

CALIBRATION OF SOIL TESTS

Soil tests as an aid for estimating available micronutrients have been used for many years. Such tests are empirical and must be correlated with crop response on a large number of soils ranging from low to high levels of available micronutrient before they can be used for diagnostic purposes. Initially, such studies are often done in the greenhouse, where it is possible to compare a large number of soils with varying chemical and physical properties under the same controlled growing conditions. Several different crops can be grown simultaneously in the greenhouse; this may not be possible under field conditions. Greenhouse studies require less time and money than do field studies.

Some limitations of greenhouse studies are as follows: (i) the restricted root volume in greenhouse pots can accentuate crop response to applied nutrients, (ii) poor light quantity and quality as well as lack of optimum amounts of other nutrients may influence crop response and yields, (iii) subsoil and climatic effects cannot be readily measured, and (iv) some crops cannot be successfully grown to maturity in the greenhouse. Field studies are thus required for calibration of micronutrient soil test procedures, just as they are for macro- and secondary nutrients.

Field calibration of micronutrient soil tests with crop responses are generally more difficult than those for P and K. This is due partly to the difficulty of finding sufficient field sites where crops will respond to the applied micronutrients. Proper field calibration studies require that experiments be conducted on various soils having the full range of available micronutrients.

Contamination is always a problem in micronutrient studies. Small changes in micronutrient concentrations due to contamination cause relatively large errors in analytical results because micronutrient concentrations in plant tissue and soil extracts are usually low (< 200 ppm). These errors frequently result in misinterpretation of the data. Contamination of plant samples is more difficult to prevent under field than greenhouse conditions. Dust and other aerial contaminants in the field cannot always be removed completely from plant tissue. However, every effort must be made to reduce such possible contamination with field samples.

Approaches Used in Calibration

Several approaches have been used to calibrate micronutrient soil tests. Each approach has its limitations; so combinations of all three approaches may lead to more effective calibrations.

YIELD RESPONSE

Crop growth response to the applied micronutrient is perhaps the best approach. Increased crop growth resulting from a micronutrient application verifies that the soil was deficient in that micronutrient, at least under the conditions of that particular growth situation. The concentration of chemically extractable micronutrients which separates deficient from nondeficient soils is called the critical soil level. Trierweiler and Lindsay (1969) used this approach to calibrate their EDTA-$(NH_4)_2CO_3$ soil test for available Zn on 42 Colorado soils. Soils testing greater than 1.4 ppm in available Zn were considered adequate for corn (*Zea mays* L.), while those with soil test levels below this critical value were classed as deficient (Fig. 1). The 0.1 *N* HCl extraction procedure was inadequate for separating deficient from nondeficient soils in this study. Its failure to provide a good separation was attributed to the variable free $CaCO_3$ contents of these soils. Neutralization of the acid in these calcareous soils reduces acid strength and can result in lower levels of extractable Zn.

Brown, Quick, and Eddings (1971) used the same approach to calibrate several extraction procedures for Zn on 92 California soils. Critical levels for soil Zn extracted by dithizone, DTPA, and EDTA were 0.55, 0.50, and 1.25 ppm, respectively. They concluded that the DTPA test developed by Lindsay and Norvell (W. L. Lindsay and W. A. Norvell, 1969. Development of a DTPA micronutrient soil test. Agron. Abstr. 69-84) was preferable to other soil tests. For some soil associations more than one method may be required to properly assess the available Zn status.

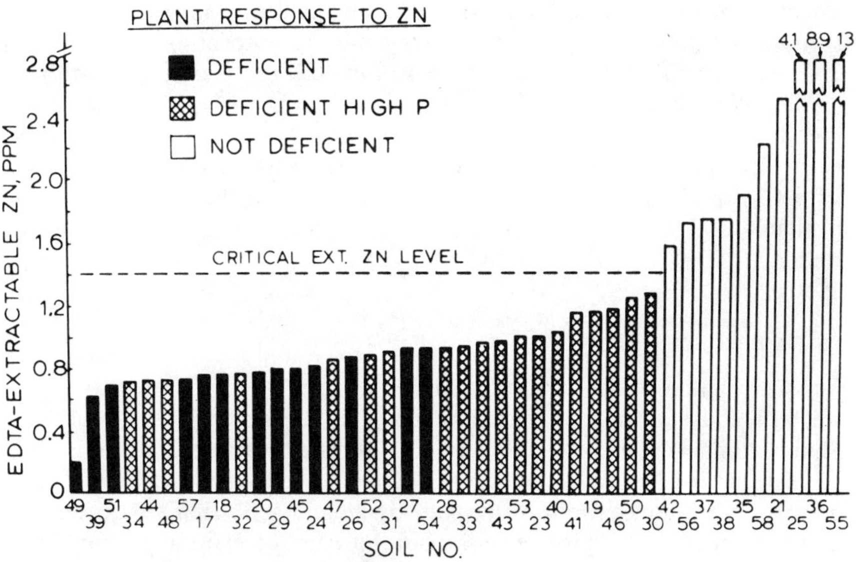

Figure 1. The EDTA-$(NH_4)_2CO_3$-extractable Zn of 42 Colorado soils in relation to Zn response of corn in the greenhouse (Trierweiler & Lindsay, 1969).

This requirement adds to the work of the soil test laboratory, and it may be preferable to change to a method which is usable over the entire range of soils encountered.

Mehlich and Bowling (1975) reported limited field results showing increased yields of wheat (*Triticum aestivum* L.) and soybeans (*Glycine max* L.) with Cu applications to soils containing less than 0.5 ppm extractable Cu. They extracted soils with 0.5 *N* HCl and 0.05 *N* $AlCl_3$. The strong acid extractant was used to obtain high enough extract concentrations to measure Cu directly by atomic absorption spectroscopy without a time-consuming concentrating procedure. The latter procedure may still be necessary if the detection limit of the instrument is too high for analysis of soils with low levels of available Cu.

NUTRIENT UPTAKE

Plant uptake of micronutrients is often used to calibrate soil tests under greenhouse conditions. Application of a micronutrient to a deficient soil generally results in increased uptake, with or without yield response; uptake is usually a more reliable criterion than nutrient concentrations for correlation studies. All nutrients other than the one(s) being tested must be present in nonlimiting amounts. If not, growing conditions will not be optimum and crop response to the nutrient under test will not be maximized.

Stewart and Berger (1965) concluded that soil Zn extracted by 2 *N* $MgCl_2$ was more closely related to Zn uptake by Japanese millet (*Echinochloa frumentacea* L.) grown on 42 Wisconsin soils than was Zn extracted by 0.1 *N* HCl or $NH_4C_2H_3O_2$-dithizone. Problems arise from high salt concentrations in extracts plugging atomic absorption burners, discouraging use of such extractants for routine soil test procedures. Martens (1968), Wear and Evans (1968), and Evans et al. (1974) also related the amount of extractable Zn in soils to Zn uptake. Blevins and Massey (1959) used this approach to correlate soil Cu extracted by EDTA or dithizone with Cu uptake by millet; they did not recommend these extractants for available soil Cu because of the relatively low correlation coefficients obtained.

NUTRIENT CONCENTRATIONS

Correlation of the concentration of extractable micronutrients in the soil with concentrations in plant tissue is often used in calibrating soil tests. Shuman and Anderson (1974) treated eight southeastern U.S. soils with several rates of Mn and grew wheat and soybeans in the greenhouse. They correlated Mn concentrations in plant tissue and soil using six extractants and concluded that DTPA gave the best measure of plant available Mn in most of the soils studied. However, their data showed that several other chelates and inorganic extractants, including the double acid extractant

($0.05\ N$ HCl + $0.025\ N\ H_2SO_4$) used for available P and K in this region, also gave highly significant correlations. Hoff and Mederski (1956) used the same approach to show soil Mn extracted by $0.1\ N\ H_3PO_4$ or $3\ N\ NH_4H_2PO_4$ was highly correlated with Mn concentrations in soybeans. Haq and Miller (1972) found that EDTA- and DTPA-extractable Zn in 86 Canadian soils reflected the concentration of Zn in corn grown on these soils under greenhouse conditions. These extractants were superior to EDDHA and the double acid.

One assumption made in this approach is that nutrient concentrations in plants are directly related to the level of available nutrient in the soil. Nutrient concentrations in plants are also related to growing conditions, amount of growth, and growth stage. If the rate of plant growth is reduced by low levels of nutrients other than the one under test, concentrations in the plant tissue can be higher than expected. Nutrient dilution resulting from increased growth rates also occurs when the limiting factor has been removed (Terman et al., 1975). Thus, nutrient concentration and uptake, at least during early growth, should be evaluated in terms of growth rates and total dry matter production.

Relationships with Other Soil Properties

In the development of soil tests, many investigators have attempted to relate the concentration of a particular micronutrient in various soil extracts to other soil properties, such as soil pH, organic matter content, CEC, percent clay (a measure of soil texture), and available levels of other nutrients. Reactions of a particular micronutrient with soil are closely related to the above properties, and the extent of these reactions should affect the level of extractable micronutrients to some degree. In other cases the extractable micronutrient concentration may not be significantly related to plant uptake unless other factors are considered. Inclusion of the effects of these soil variables should be a necessary part of the development of soil tests.

Martens (1968) used multiple regression analyses to show that B uptake by corn was best predicted by hot H_2O-soluble B and soil pH, while Cu uptake was best predicted by inclusion of soil organic matter content with the level of extractable Cu. The predictions of Zn uptake were not improved by inclusion of soil pH, organic matter content, or percent clay. In contrast, Osiname, Schulte and Corey (1973) showed the best prediction of Zn uptake by oats (*Avena sativa* L.) from 28 soils in western Nigeria resulted when the above factors were included in the regression analyses. Uptake of Cu was more strongly correlated with the inorganic soil fractions than with organic matter content. A significant amount of the available Zn and Cu is retained by the sesquioxides in highly weathered soils, whereas more of the available Zn, and especially Cu, are associated with the organic fraction in less weathered temperate soils.

Dolar and Keeney (1971) related plant uptake to extractable Cu, Mn, and Zn in 36 Wisconsin soils in a greenhouse study. They found that 0.01 *M* EDTA in *N* $NH_4C_2H_3O_2$ (pH 7.0) or 0.1 *N* HCl could be used for simultaneous determination of available Cu, Mn, and Zn. Correlation and multiple regression analyses showed that Cu uptake by oats was best predicted by inclusion of soil pH or extractable Mn and Zn; Mn uptake by inclusion of available P, extractable Cu, or Zn uptake; and Zn uptake by inclusion of soil organic matter content and available P, extractable Cu and Mn, or Mn uptake in the regression equations. Results of this study suggest a marked interaction of these three micronutrients in plant nutrition. Inclusion of these parameters in regression equations for interpretation of routine soil analyses may require chemical analyses normally not included in routine soil tests. The resulting increased cost of extra analyses usually rules out such soil tests if more practical tests can be developed and correlated with plant response.

Measurement of soil pH is always included in soil tests, so this parameter can be included in interpretation without extra laboratory work. Haq and Miller (1972) included the double acid, EDDHA, EDTA, and DTPA extractants in a calibration study for determination of available Zn in 85 Ontario soils. Soil Zn extracted by DTPA or EDTA was more closely correlated to plant uptake of Zn than by the other two extractants. Results showed that if these extractants are to be used for acid soils varying in pH, the pH must be included in a regression equation to obtain a satisfactory interpretation of available Zn.

Both the EDTA and DTPA extractants were developed for neutral and calcareous soils. The EDTA extractant is buffered at pH 8.6 and DTPA is buffered at pH 7.3. Therefore, small differences in soil pH in neutral and calcareous soils will not affect the level of extractable Zn to the degree that occurs with other extractants. Nelson, Boawn, and Viets (1959) found that the 0.1 *N* HCl extractant could be used for estimating available Zn on calcareous soils if proper adjustment was made for the titratable alkalinity (a measure of free $CaCO_3$) in the soil. This method requires another measurement on each soil. The DTPA extractant has now replaced the acids for determination of available Zn in states with large hectarages of calcareous soils.

Relationships between soil pH and available Mn are especially important. Cox (1968) developed a soil test interpretation which was applicable over a soil pH range of 5.2 to 7.1 and a double acid-extractable Mn range of 1 to 10 ppm. The level of extractable Mn required for normal growth increases with soil pH. Shuman and Anderson (1974) reported that H_2O-soluble and exchangeable Mn of extremely acid soils were better related to plant uptake of Mn, while DTPA- and double acid-extractable Mn gave better correlations in soils with pH values > 5.8.

Randall, Schulte, and Corey (1976) reported significant correlations between soil Mn recovered by numerous extractants and Mn uptake by oats followed by ryegrass (*Lolium perenne* L.) in a greenhouse study. The inclusion of soil pH and organic matter content improved Mn uptake predictability in regression equations on soils with low (< 6%) organic matter. Soil organic matter content and the pH × organic matter cross-product were more important than the soil test value in high (> 6%) organic matter soils; these accounted for 70% of the variation in Mn uptake. The R^2 values were increased to 80% when soil Mn extracted by DTPA or EDTA-$(NH_4)_2CO_3$ was included in the equations. These results show that plant availability of Mn is closely related to soil pH and that soil organic matter content should also be considered in Mn soil test interpretation.

Regional Calibration Studies

In 1971 a Southeast Regional Micronutrient Technical Committee was formed to develop and/or calibrate soil tests for available Zn. Representatives conducted similar calibration studies in seven southeastern states with corn and in three states with rice (*Oryza sativa* L.). All soil samples were sent to the same laboratory for chemical analyses. Three extractants, 0.1 *N* HCl, double acid, and DTPA, were compared in this study.

Preliminary results show that all three extractants can be used for determination of available Zn in this region. States which use the double acid extraction for available P and K will undoubtedly use this same extract for determination of available Zn. Other states not using this extractant for available P may use 0.1 *N* HCl for available Zn. The DTPA procedure probably will not be used in this region, as it is more time consuming than the other two procedures, and it does not give better correlations than the two acid extractants on noncalcareous soils (F. R. Cox, Technical Committee Chairman, personal communication).

Other regional studies may help in the calibration of micronutrient soil tests. This may result in fewer soil test methods and critical soil levels reported for crops on similar soils within a given region rather than on a state level.

INTERPRETING SOIL TEST RESULTS

Results of soil tests must be translated into a fertilizer recommendation to be of use to the grower. Suitable fertilizer recommendations must include several factors in addition to the soil test value. Micronutrient requirements of different crop species vary greatly, so it is necessary to know the crop to be grown. For example, corn is more Zn-sensitive than grain sorghum (*Sorghum bicolor* L. Moench) but the reverse is true for Fe. Critical levels for Zn and Fe vary with these two crops. Cultivars of the same crop species also differ significantly in their uptake and utilization of micronutrients. Use of cultivars tolerant to marginal deficiencies of some micronutrients, especially Fe, is the recommended practice rather than fertilizer applications.

Rates of micronutrient fertilizers must also be related to the desired yield level and potential economic return of the crop planted. Results of experiments should give the probable yield responses at given soil test levels, from which the probability of a profitable response to micronutrient fertilization can be estimated. The previous cropping history as well as the effects of weather on crop response to micronutrient fertilizers should be considered when making the fertilizer recommendation. Micronutrient fertilizer recommendations for many crops were reviewed at length by Murphy and Walsh (1972).

The philosophy of micronutrient fertilization also affects the fertilizer recommendation. Some suggest that micronutrients be applied to insure a sufficient supply of available micronutrients to the crop, especially when the soil test level is marginal. Growers are more apt to apply micronutrients to high cash value crops such as vegetables to insure optimum soil fertility. Relative differences between cost of fertilizers and the decreased income from possible yield losses due to an inadequate supply of micronutrients are much greater with such crops.

For most agronomic crops, micronutrient fertilizers are not recommended unless results of experiments have shown that positive yield responses can be obtained when micronutrient availability is below the critical soil test level. Probability of unfavorable weather should also be considered. For example, Zn may be recommended on soils with marginal levels of available Zn for some crops when cool, wet weather is expected to occur, because plant availability of soil Zn may be decreased under these conditions.

Reporting Results

Some micronutrient soil test reports give the general range of micronutrient availability; others give the concentration of extractable micronutrient as well. Most growers are interested in both the general level of available micronutrient and the fertilizer recommendation. Actual concentrations of extractable micronutrients are of little value without interpretation and recommendations.

Three general levels of micronutrient availability are used by some laboratories in reporting results and making fertilizer recommendations. This allows a general differentiation in probable response of highly susceptible and relatively tolerant crops to a given micronutrient. Soil test levels for DTPA-extractable Fe and Zn in the Great Plains states are classified as low, medium (marginal), and high (Table 1). Most crops respond to the recommended micronutrient on soils testing low in Fe or Zn under good management. However, only highly susceptible crops may respond to the applied micronutrient on soils in the medium range.

Table 1. Interpretive guide for DTPA-extractable Fe and Zn in the Great Plains states

Micronutrient	Low	Medium	High
		— ppm—	
Fe	0-2.5	2.5-4.5	>4.5
Zn	0-0.5	0.5-1.0	>1.0

Table 2. Interpretive guide for DTPA-extractable Zn for several crops on California soils (Reisenauer & Quick, 1971)

Crop*	Highly responsive	Probably responsive	Not responsive
		— ppm —	
Barley and Wheat	0-0.2	0.2-0.3	>0.3
Corn	0-0.3	0.3-0.6	>0.6
Cotton	0-0.4	0.4-0.7	>0.7
Potatoes	0-0.3	0.3-0.7	>0.7
Sugar beets	0-0.1	0.1-0.2	>0.2
Vegetables (cool season)	0-0.5	0.5-1.0	>1.0

* Botanical names of crops not listed in the text are: barley (*Hordeum vulgare* L.); cotton (*Gossypium* sp.), and potatoes (*Solanum tuberosum*).

Micronutrient fertilizer would not be recommended for any crop on soils having test results interpreted as high.

In California the three classifications shown in Table 2 are used to interpret soil test levels of available Zn. In addition, the concentration range of DTPA-extractable Zn in each category is changed according to the susceptibility of the crop to Zn deficiency. The least Zn-sensitive crop, sugar beets (*Beta vulgaris* L.), has the lowest critical value (0.2 ppm), of those listed in Table 2. This interpretive guide is more specific than most systems and requires more calibration studies for implementation. It includes the medium category, which allows for judgements of other factors in making fertilizer recommendations.

In North Carolina the concentration of an extractable micronutrient in the soil is converted to an index number by relation to the critical soil level; fertilizer recommendations are made accordingly. Hatfield (1972) suggested this system as an attempt to provide a common language that would accommodate various soil test procedures. This system relates each laboratory's evaluation of soil test levels to predicted sufficiency levels of each nutrient. Crop response to a nutrient is correlated to soil test levels and the values are fitted to a 0-100 index from deficiency to optimum levels. In Mehlich and Bowling's method (1975) concentrations of extractable Cu are converted to index numbers and Cu applications are recommended on soils with a Cu index less than 10 (0.5 ppm extractable Cu).

Recommended rates are 2 to 3 and 3 to 5 kg/ha of Cu for mineral and organic soils, respectively.

Recommendations for Mn are now made for two crops in North Carolina using the index system. Critical soil pH values are predicted, above which crop response to Mn would be probable. The following equations are used for determination of the predicted soil pH (Y) from the Mn index (I):

$$\text{Soybeans } Y = 5.1 + 0.0458(I) \quad [1]$$

$$\text{Peanuts } Y = 5.8 + 0.0207(I). \quad [2]$$

For example, a soil testing 2 ppm of double acid-extractable Mn would have an index number of 10. The predicted critical pH for soils with an index number of 10 would be 5.6 and 6.0 for soybeans and peanuts (*Arachis hypogaea* L.), respectively. Manganese fertilizer or foliar sprays would be recommended if the measured soil pH was higher than the predicted soil pH. This system allows the inclusion of a second parameter, soil pH, in the interpretation of the Mn soil test (F. R. Cox, North Carolina State University, personal communication).

Effects of Extraction Conditions on Interpretations

During the development of a soil test, the investigators generally study the effects of changing the soil/solution ratio, extraction time, and extractant concentration on the level of extractable micronutrient. The extraction conditions which give soil test values that are best correlated with plant response are chosen. However, extraction conditions must be adaptable for routine analysis if the soil test is to be practical.

An examination of published soil test procedures shows a wide variety of extraction conditions. Many of these conditions are similar but do not agree in all details. As a result, the critical soil values for a given micronutrient may differ from one laboratory to another for the same crop on similar soils. For example, the 0.1 *N* HCl procedure is widely used for determination of available Zn. Wear and Sommer (1948) first reported the feasibility of this soil test. They extracted 10 g of soil with 100 ml of 0.1 *N* HCl for 60 minutes. Others have reported a range from 1:4 to 1:10 in soil/solution ratios, and 5 to 60 minutes of extraction time. Effects of soil pH, texture, organic matter content, and other parameters have resulted in changing the extraction parameters for the best correlation. Yet the soil test in most instances is called the 0.1 *N* HCl-extractable Zn procedure.

Unless otherwise defined, an extraction procedure implies that the extraction conditions are constant. It appears that there would be less confusion regarding soil test levels in the literature if "standard" extraction conditions for the most popular procedures were used. Then micronutrient soil test results would be more comparable. The critical levels for a given crop would, of course, depend on the results of calibration studies.

Even slight changes in laboratory procedures may affect the amount of micronutrient extracted from soils. Interpretation of the results could change if these effects are relatively large. A small change in extraction or filtration time may affect the results, particularly if the extraction time is short. Effects of changing other parameters may become more important if the extraction time is long, as with the 2-hour extraction for the DTPA soil test.

Soil Test Laboratory personnel at Colorado State University noted in 1974 that the percentage of soil samples testing low in available Fe had increased. They investigated the effects of several changes in the DTPA procedure which had taken place. A new shaker and different shaped extraction flasks had been installed in their laboratory. Thirty-five soil samples were then extracted on both new and old shaker setups. They found that the level of DTPA-extractable Fe in these soils was lower with the new, less vigorous shaking conditions. The critical level for available Fe decreased from 4.5 to 3.3 ppm. Khan and Soltanpour showed that changes in the shape of extraction flask and oscillation speed of the shaker affected the level of DTPA-extractable Fe and Zn in a number of soils (A. Khan and P. N. Soltanpour, 1974. Effect of type and speed of shaker and type of utensil of DTPA-extractable Fe, Zn, and Mn in soil. Western Soil Sci. Soc. Abstracts. Fort Collins, Colorado).

Soil test and fertilizer recommendation bulletins of at least four Great Plains states (Colorado, Kansas, Nebraska, and Oklahoma) contain information regarding the DTPA test for extractable Fe. The critical soil level for grain sorghum is listed as 4.5 ppm in these bulletins. All of these soil test laboratories do not have the same type of extraction flasks and shaking equipment. Representatives from each of these laboratories will exchange soil samples to determine the proper extraction conditions required to obtain the same level of DTPA-extractable Fe and Zn in each laboratory (P. N. Soltanpour, Colorado State University, personal communication). Critical soil levels for Fe and Zn will thus be the same for similar soils tested by these laboratories.

SOIL TESTS FOR RESIDUAL EFFECTS OF MICRONUTRIENTS

Continued annual applications of micronutrients to most soils will result in a buildup of these nutrients. Crop recovery during the first year after application is generally 5 to 10% of the applied micronutrient. Mobility of most micronutrients in soils is quite low, so little of the micronutrient fertilizers will be leached from the surface soil. Thus, some of the applied micronutrients should remain in the root zone in a form available to plants.

Boawn et al. (1960), reported a significant residual effect after 5 years with $ZnSO_4$ applications of 9 or 18 kg/ha of Zn to a neutral, noncalcareous soil in Washington. The amount of 0.1 *N* HCl-extractable Zn declined with time but crop removal of applied Zn during this period accounted for only about 13% of this decline. Vinande et al. (1968), in Michigan also noted a significant residual effect on pea beans (*Phaseolus* sp.) from 28 kg/ha of Zn as $ZnSO_4$ applied 3 years previously. Schnappinger et al. (1972), in Virginia reported greater yield increases from residual applications of 7 and 14 kg Zn/ha as $ZnSO_4$ than from 4.4 kg Zn/ha as ZnEDTA. Residual value of various micronutrient sources can thus be different depending upon their rates of reversion to less soluble forms in the soil.

Soil extractants can be used to determine buildup of some micronutrients in soils following periodic additions to a soil. Martens, Carter, and Jones (1974) found that soybean plant growth and seed yield were not affected with rates up to 3.3 kg B, 8.4 kg Cu, and 11.1 kg Zn/ha applied annually for 6 years. Concentrations of hot H_2O-soluble B and 0.1 *N* HCl-extractable Cu and Zn were significantly increased as a result of these applications. Follett and Lindsay (1971) followed changes in DTPA-extractable Cu, Fe, Mn, and Zn in 11 Colorado soils after fertilization. Decreases in extractable levels with time were related to residual response from these micronutrients, suggesting that the DTPA soil test may be useful in monitoring the availability of these micronutrients in fertilized as well as unfertilized soils. Boawn (1974) showed that a $ZnSO_4$ application of 5.6 kg Zn/ha was adequate for 2 years, while a Zn rate of 11.2 kg/ha was adequate for 4 years. Results were interpreted in terms of 0.8 ppm of DTPA-extractable Zn as the critical level.

Similar studies are needed for all micronutrients on various other soil types. Results of these studies may show that there is a marked residual response to some micronutrients. It may be easier to apply a larger micronutrient rate once every 3 to 5 years than small annual applications for some crops. Soil test values could be used to show when levels of available micronutrients have decreased to the point where fertilizers would be recommended again. Care must be taken with this approach in fertilization of B, and possibly Cu, as the range between deficiency and toxicity of these nutrients is rather narrow in some soils or for some sensitive crops.

MONITORING EFFECTS OF OTHER SOIL APPLICATIONS

Soil tests also may be of value in detecting excesses of available micronutrients. While micronutrient deficiencies are much more prevalent than toxicities, the latter may occur occasionally. Crop toxicities are more likely to result from increased availability of indigenous micronutrients with increased soil acidity (except for Mo) than to overapplication of micronutrients. Boawn and Rasmussen (1971) applied $ZnSO_4$ at rates up to 500 ppm of Zn to an alkaline soil (pH 7.5) and found significant yield decreases on 12 of 18 plant species. Concentrations of extractable Zn in soil treated with 10 and 500 ppm of Zn were 11 and 450 ppm with 0.1 *N* HCl and 5 and 246 ppm with DTPA. Thus, both extractants could be used to measure large increases in available Zn level in this soil. Baker and Tucker collected soil samples around a Zn smelter plant and found concentrations as high as 450 ppm of 1.0 *N* HCl-extractable Zn (J. M. Baker and B. B. Tucker, 1972. Toxic soil levels of zinc and cadmium. Agron. Abstr. 72-103). Concentrations of DTPA-extractable Zn were about half as high. Plant growth of wheat and sorghum decreased with increasing Zn levels in this soil. Vlek and Lindsay reported that 1.0 *M* $(NH_4)_2CO_3$-extractable Mo was highly correlated with Mo in alfalfa (*Medicago sativa* L.) in soils > pH 7.0 where Mo toxicity is a problem (P. L. G. Vlek and W. L. Lindsay, 1975. Soil tests for determining molybdenum contamination in soils. Agron. Abstr. 75-122). The DTPA extractant also was adequate, but analysis of Mo in this extractant was difficult, resulting in poor reproducibility.

Soil application of municipal and industrial wastes and use of agricultural chemicals containing micronutrients may also increase the soil test levels. Therefore, it is important to monitor the status of available micronutrients in soils receiving these amendments.

Fly Ash Applications

Soil disposal of industrial wastes such as fly ash from coal ash residues at electric power generating plants may result in increases in levels of some plant nutrients. The B concentration of many fly ash materials is relatively high (> 200 ppm), so relatively high amounts of B are applied to soil at the usual disposal rates. Toxicity to some plant species is possible if B in the fly ash products is readily available. Plank and Martens (1974) showed that hot H_2O-soluble B extractions could be used to predict the plant availability of B in soils treated by fly ash. They suggest that certain considerations also be made in determining ash disposal rates of fly ash. These include soluble salt damage to plants, toxic levels of some elements in the product, and possible micronutrient deficiencies due to the liming effect of fly ash. Further studies are required to determine possible agronomic effects of this practice; soil test results may help monitor these effects.

Sewage Sludge Applications

High application rates of sewage sludge can result in relatively high applications of micronutrients and other heavy metals to soils. Results of a greenhouse experiment showed that growth of corn was not affected by a sewage sludge application rate equivalent to 224 metric tons/ha (Mortvedt & Giordano, 1975). Amounts of Zn and Cr contained in the sludge were each equivalent to 3.1 metric tons/ha. Levels of 0.5 *N* HCl-extractable Zn and Cr in the soil were increased from 2 and 1 ppm to 635 and 430 ppm, respectively, by the sludge application. Tissue concentrations of Zn increased from 14 to 508 ppm, while those of Cr were not affected. Results also showed that 0.1 *N* HCl extracted almost the same amount of Zn but much lower amounts of Cr than those extracted by 0.5 *N* HCl.

Pesticide Applications

Walsh et al. (1972) applied $ZnSO_4$ to a loamy sand at rates up to 363 kg/ha to approximate the amount of Zn applied with up to 81 annual fungicidal treatments. Yields of snapbeans (*Phaseolus vulgaris* L.), corn, and cucumbers (*Cucumis sativus* L.) were not reduced by Zn applications. Highly significant correlations were observed between the Zn extracted by 0.1 *N* HCl, EDTA, or DTPA and Zn uptake of several crops. Values as high as 47 ppm extractable Zn were noted at the high $ZnSO_4$ application rate. Fiskell and Westgate (1955) reported increased levels of available Cu in soils where Bordeaux mixture had been used for many years as a fungicide for vegetables. Decreased crop production due to Cu-induced Fe chlorosis was evident on some fields. Levels as high as 2.1 ppm of H_2O-soluble CU were reported. Vitosh, Warncke, and Lucas (1973) warn growers in Michigan of potential Cu toxicity on soils treated repeatedly with Cu in fertilizers or as fungicides.

Research Needs

These results point out the need for monitoring levels of available micronutrients in soils which may be approaching the toxic range. More studies such as those discussed above are needed to determine potentially toxic micronutrient levels by routine soil tests. When these relationships are determined it should be possible to quickly diagnose possible toxicities due to micronutrient buildup.

Investigators may find that extraction procedures other than the common ones give better correlations in the toxic range, but use of a special procedure would require extra analyses. Therefore, it should be preferable to correlate soil test data from routine analyses with crop response in the toxic range as done by Walsh et al. (1972). Some soil test laboratories now use five categories to report their data: very low, low, medium, high, and very high. The latter classification would indicate a potentially toxic condition which should be examined carefully. Similar relationships for heavy metals should be studied, preferably with procedures already in use in soil test laboratories.

SUMMARY

Soil tests can be useful in diagnosing micronutrient deficiencies. Concentrations of micronutrients in soil determined by various extractants are of little value unless they have been calibrated in terms of crop response, which requires greenhouse and field studies. The soil test data must then be interpreted in terms of fertilizer recommendations for the grower.

Several new micronutrient soil tests have been developed in recent years. For various reasons they have not always been correlated with crop response on other soil types. An examination of methods reported in the literature shows that extraction procedures vary somewhat from one laboratory to another. Because extractions for most of these procedures are nonequilibrium conditions, the level of extractable micronutrients may vary with slight changes in procedures. In some instances, selection of soil test procedures is also affected by the equipment available in the laboratory. Adherence to "standard" procedures for the well known micronutrient soil tests should minimize the discrepancies in soil test values. More regional cooperative studies would also result in establishment of critical soil levels which are applicable over larger areas.

A recent trend in micronutrient soil tests is to use the same extractant for more than one element. The double-acid procedure is now used for available Zn and Mn as well as for P, K, Mg, and Ca in some states where acid soils predominate. Acid extractants are not accurate for calcareous soils; instead, several chelate extractants are now used for micronutrient cations in these soils. Available B is still best measured by the hot water procedure. Available Cu and possibly Mn are related to soil organic matter content, so specific soil test procedures may be required. As available Mn is closely related to soil pH, the latter should be considered in Mn soil test interpretations. It would appear preferable for those involved in this research area to agree on the use of fewer soil extractants and standardized procedures. Critical soil levels obtained for various soil and crop conditions may then have wider application than at present. Small modifications in the above procedures will only change the critical levels that are already in use. Some reported critical soil levels for various crops as determined with various extractants are shown in Table 3.

Proper interpretation requires that several factors be considered, including previous cropping history and crop to be grown. Critical soil levels differ among crop species. Differential cultivar response to some micronutrients is well demonstrated, and use of tolerant cultivars may alleviate the need for applying micronutrients to soils with marginal deficiencies. Environmental factors such as soil temperature, moisture, and aeration markedly affect the availability of micronutrients. Lastly, the complex reactions of micronutrient sources in soils alter their effectiveness. All of these factors must be considered in making fertilizer recommendations.

Table 3. Crop response to micronutrients as related to critical soil levels obtained with various extractants

Micronutrient	Crop*	Soil test level: Response	Soil test level: No Response	Extractant	Reference
		ppm	ppm		
Boron	Clover (seed)	<0.10	--	Hot H_2O	Johnson and Wear (1967)
	Cauliflower	0.50	--	Hot H_2O	Gupta (1971)
	Peanuts	<0.15	> 0.15	Hot H_2O	Hill and Morrill (1974)
Copper	--	<9.00	>20.00	1.0 *N* HCl	Vitosh et al. (1973)
	Oats	0.50	1.20	0.5 *N* HCl-$AlCl_3$	Mehlich and Bowling (1975)
Iron	Corn	0–2.0	> 2.10	DTPA	Whitney et al. (1973)
	Sorghum	0–4.5	> 4.50	DTPA	Whitney et al. (1973)
Manganese	Soybeans	<20	--	0.1 *N* H_3PO_4	Hoff and Mederski (1956)
	Oats	<20	--	0.1 *N* H_3PO_4	Hammes and Berger (1960)
	--	<19†	--	Double acid	Mehlich (1957)
	Oats	<10†	--	Double acid	Cox (1968)
	--	< 5†	>20†	0.1 *N* HCl	Vitosh et al. (1973)
Molybdenum	Corn	0.09	--	$(NH_4)_2C_2O_4$	Hagstrom and Berger (1965)
	Clover	0.16	--	$(NH_4)_2C_2O_4$	Hagstrom and Berger (1965)
	Peas	0.42	--	$(NH_4)_2C_2O_4$	Hagstrom and Berger (1965)
Zinc	Corn	<1.40	--	EDTA-$(NH_4)_2CO_3$	Trierweiler and Lindsay (1969)
	Corn	<0.50	--	DTPA	Brown et al. (1971)
	--	<2.50†	>10†	0.1 *N* HCl	Vitosh et al. (1973)
	Corn	<0.50	0.5	DTPA	Whitney et al. (1973)

* Botanical names of crops not listed in the text are: clover (*Trifolium* sp.); cauliflower (*Brassica oleracea* var. *botrytis*); and peas (*Pisum sativum*).
† Probability of crop response at a given soil test level also related to soil pH.

Soil tests may have value in showing the residual effects of micronutrient fertilizers applied to soils. First year crop recovery of applied micronutrients is much lower than that of P and K so residual value of some micronutrient sources may be considerable. If correctly calibrated, soil test results could be used to show the changing micronutrient availability status in soils.

Soil tests also may be of use to detect potential toxicities of some micronutrients. While toxicities are not extensive, a calibrated soil test to assess the excess range would be of value. If soils are to be used for disposal of municipal and industrial wastes, knowledge of potential toxicities due to buildup of some micronutrients and other heavy metals as Cd, Ni, and Pb will be invaluable.

Development of soil tests is more complex for micronutrients than for P and K. Recent improvements in analytical methods as well as a better understanding of micronutrient chemistry in soils and plant nutrition have led to improvements in micronutrient soil tests. The use of soil tests to predict probable response to micronutrient fertilization must be refined further so that more confidence can be placed in this diagnostic tool.

LITERATURE CITED

Blevins, R. L., and H. F. Massey. 1959. Evaulation of two methods of measuring available soil Cu and the effects of pH and extractable aluminum on copper uptake by plants. Soil Sci. Soc. Am. Proc. 23:296-298.

Boawn, L. C. 1974. Residual availability of fertilizer zinc. Soil Sci. Soc. Am. Proc. 38:800-804.

Boawn, L. C., and P. E. Rasmussen. 1971. Crop response to excessive zinc fertilization of alkaline soil. Agron. J. 63:874-876.

Boawn, L. C., F. G. Viets, Jr., C. L. Crawford, and J. L. Nelson. 1960. The effect of nitrogen carrier, nitrogen rate, zinc rate, and soil pH on zinc uptake by sorghum, potatoes, and sugar beets. Soil Sci. 90:329-337.

Brown, A. L., J. Quick, and J. L. Eddings. 1971. A comparison of analytical methods for soil zinc. Soil Sci. Soc. Am. Proc. 35:105-107.

Cox, F. R. 1968. Development of a yield response prediction and manganese soil test interpretation for soybeans. Agron. J. 60:521-524.

Cox, F. R., and E. J. Kamprath. 1972. Micronutrient soil tests. p. 289-317. *In* J. J. Mortvedt, P. M. Giordano, and W. L. Lindsay (ed.) Micronutrients in agriculture. Soil Sci. Soc. of Am., Madison, Wis.

Dolar, S. G., and D. R. Keeney. 1971. Availability of Cu, Zn, and Mn in soils. I. Influence of soil pH, organic matter, and extractable phosphorus. J. Sci. Food Agric. 22:273-278.

Evans, C. E., J. I. Wear, B. F. Hajek, and J. T. Cope, Jr. 1974. The relationship of soil zinc removed by three extractants to zinc uptake by corn and sorghum in medium- to fine-textured soils. Commun. Soil Sci. Plant Anal. 5(2):105-113.

Fiskell, J. G. A., and P. J. Westgate. 1955. Copper availability of high copper soils. Proc. Fla. State Hortic. Soc. 68:192-197.

Follett, R. F., and W. L. Lindsay. 1971. Changes in DTPA-extractable zinc, iron, manganese, and copper in soils following fertilization. Soil Sci. Soc. Am. Proc. 35:600-602.

Gupta, U. C. 1971. Boron requirement of alfalfa, red clover, brussel sprouts, and cauliflower grown under greenhouse conditions. Soil Sci. 112:280-281.

Hagstrom, G. R., and K. C. Berger. 1965. Molybdenum deficiencies of Wisconsin soils. Soil Sci. 100:52-56.

Hammes, J. K., and K. C. Berger. 1960. Manganese deficiency in oats and correlation of plant manganese with various soil tests. Soil Sci. 90:239-244.

Haq, A. U., and M. H. Miller. 1972. Prediction of available soil Zn, Cu, and Mn using chemical extractants. Agron. J. 64:779-782.

Hatfield, A. L. 1972. Soil test reporting—a nutrient index system. Commun. Soil Sci. Plant Anal. 3:425-436.

Hill, W. E., and L. G. Morrill. 1974. Assessing boron needs for improving peanut yield and quality. Soil Sci. Soc. Am. Proc. 38:791-794.

Hoff, D. J., and H. J. Mederski. 1956. The chemical estimation of plant available soil manganese. Soil Sci. Soc. Am. Proc. 20:129-132.

Johnson, W. C., and J. I. Wear. 1967. Effect of boron on white clover seed production. Agron. J. 59:205-206.

Martens, D. C. 1968. Plant availability of extractable boron, copper, and zinc as related to selected soil properties. Soil Sci. 106:23-28.

Martens, D. C., M. T. Carter, and G. D. Jones. 1974. Response of soybeans following six annual applications of various levels of boron, copper, and zinc. Agron. J. 66:82-84.

Mehlich, A. 1957. Aluminum, iron and pH in relation to lime induced manganese deficiencies. Soil Sci. Soc. Am. Proc. 21:625-628.

Mehlich, A., and S. S. Bowling. 1975. Advances in soil test methods for copper by atomic absorption spectrophotometry. Commun. Soil Sci. Plant Anal. 6:113-128.

Mortvedt, J. J., and P. M. Giordano. 1975. Response of corn to zinc and chromium in municipal wastes applied to soil. J. Environ. Qual. 4:170-174.

Murphy, L. S., and L. M. Walsh. 1972. Correction of micronutrient deficiencies with fertilizers, p. 347-387. *In* J. J. Mortvedt, P. M. Giordano, and W. L. Lindsay (ed.) Micronutrients in agriculture. Soil Sci. Soc. of Am., Madison, Wis.

Nelson, J. L., L. C. Boawn, and F. G. Viets, Jr. 1959. A method for assessing zinc status of soils using acid-extractable zinc and "titratable alkalinity" values. Soil Sci. 88:275-283.

Osiname, O. A., E. E. Schulte, and R. B. Corey. 1973. Soil tests for available copper and zinc in soil in western Nigeria. J. Sci. Food Agric. 24:1341-1349.

Plank, C. O., and D. C. Martens. 1974. Boron availability as influenced by application of fly ash to soil. Soil Sci. Soc. Am. Proc. 38:974-977.

Randall, G. W., E. E. Schulte, and R. B. Corey. 1976. Correlation of plant available manganese with extractable soil manganese and soil factors. Soil Sci. Soc. Am. Proc. 40:282-287.

Reisenauer, H. M. 1967. Availability assays for the secondary and micronutrient anions. p. 71-102. *In* G. W. Hardy (ed.) Soil testing and plant analyses, Part I Soil Testing. Soil Sci. Soc. of Am., Madison, Wis.

Reisenauer, H. M., and J. Quick. 1971. Soil test interpretive guides. p. 96-102. *In* Proceedings Statewide Conference on Soil and Tissue Testing, University of California, Davis, Calif.

Reisenauer, H. M., L. M. Walsh, and R. G. Hoeft. 1973. Testing soils for sulphur, boron, molybdenum, and chlorine. p. 173-200. *In* L. M. Walsh, and J. D. Beaton (ed.) Soil testing and plant analyses. Soil Sci. Soc. of Am., Madison, Wis.

Schappinger, M. G., Jr., D. C. Martens, G. W. Hawkins, D. F. Amos, and G. D. McCart. 1972. Response of corn to residual and applied zinc as $ZnSO_4$ and ZnEDTA in field investigations. Agron. J. 64:64-66.

Schuman, L. M., and O. E. Anderson. 1974. Evaluation of six extractants for their ability to predict manganese concentrations in wheat and soybeans. Soil Sci. Soc. Am. Proc. 38:788-790.

Stewart, J. A., and K. C. Berger. 1965. Estimation of available zinc using magnesium chloride as extractant. Soil Sci. 100:244-250.

Terman, G. L., P. M. Giordano, and N. Christenson. 1975. Corn yield effects on phosphorus, manganese, and zinc absorption. Agron. J. 67:182-184.

Trierweiler, J. R., and W. L. Lindsay. 1969. EDTA-ammonium carbonate soil test for zinc. Soil Sci. Soc. Am. Proc. 33:49-54.

Viets, F. G., Jr. 1967. Soil testing for micronutrient cations. p. 55-70. *In* G. W. Hardy (ed.) Soil testing and plant analysis, Part I. Soil Testing. Soil Sci. Soc. of Am., Madison, Wis.

Viets, F. G., Jr., and W. L. Lindsay. 1973. Testing soils for zinc, copper, manganese, and iron. p. 152-172. *In* L. M. Walsh, and J. D. Beaton (ed.) Soil testing and plant analysis. Soil Sci. Soc. of Am., Madison, Wis.

Vinande, R., B. Knezek, J. Davis, E. Doll, and J. Melton. 1968. Field and laboratory studies with zinc and iron fertilization of pea beans, corn, and potatoes in 1967. Michigan State Agric. Exp. Stn. Quart. Bull. 50:625-636.

Vitosh, M. L., D. D. Warncke, and R. E. Lucas. 1973. Secondary and micronutrients for vegetables and field crops. Michigan Agric. Ext. Bull. E-486. East Lansing, Mich.

Walsh, L. M., D. R. Steevens, H. D. Seibel, and G. G. Weis. 1972. Effects of high rates of zinc on several crops grown on an irrigated plainfield sand. Commun. Soil Sci. Plant Anal. 3:187-195.

Wear, J. I., and C. E. Evans. 1968. Relationship of zinc uptake by corn sorghum to soil zinc measured by three extractants. Soil Sci. Soc. Am. Proc. 32:543-546.

Wear, J. I., and A. L. Sommer. 1948. Acid extractable zinc of soils in relation to the occurrence of zinc deficiency symptoms of corn. Soil Sci. Soc. Am. Proc. 12:143-144.

Whitney, D. A., R. Ellis, Jr., L. S. Murphy, and G. Herron. 1973. Identifying and correcting zinc and iron deficiency in field crops. Kansas Coop. Ext. Ser. Leaflet L-360, Manhattan, Kansas.